Lecture Notes in Computer Science 13152

More information about this subseries at https://link.springer.com/bookseries/7407

Christophe Cérin · Depei Qian ·
Jean-Luc Gaudiot · Guangming Tan ·
Stéphane Zuckerman (Eds.)

Network and Parallel Computing

18th IFIP WG 10.3 International Conference, NPC 2021
Paris, France, November 3–5, 2021
Proceedings

 Springer

Editors
Christophe Cérin
Université Sorbonne Paris Nord, LIPN
Villetaneuse, France

Depei Qian
Beihang University
Beijing, China

Jean-Luc Gaudiot
University of California at Irvine
Irvine, CA, USA

Guangming Tan
Institute of Computing Technology
Beijing, China

Stéphane Zuckerman
ETIS Laboratory
CY Cergy Paris Université, ENSEA, CNRS
Cergy, France

ISSN 0302-9743   ISSN 1611-3349 (electronic)
Lecture Notes in Computer Science
ISBN 978-3-030-93570-2   ISBN 978-3-030-93571-9 (eBook)
https://doi.org/10.1007/978-3-030-93571-9

LNCS Sublibrary: SL1 – Theoretical Computer Science and General Issues

This Springer imprint is published by the registered company Springer Nature Switzerland AG
The registered company address is: Gewerbestrasse 11, 6330 Cham, Switzerland

Preface

Welcome to the proceedings of the eighteenth edition of the International Conference on Network and Parallel Computing (NPC 2021), held in Paris, France, during November 3–5, 2021.

First, we would like to honor the memory of Guang R. Gao, who, on top of having spent his life making substantial contributions to the field of high-performance computing in general, was a driving force behind NPC. Along with a few other people from the NPC Steering Committee, he suggested to hold NPC outside of Asia to broaden its audience, and have it occur in a European venue.

Second, we would like to thank all the committee members, from the program vice-chairs to the local organizing chairs and the reviewers. We are thinking in particular of Aziza Lounis from DNAC, Abdulhalim Dandoush from ESME Sudria, Éric Renault from ESIEE Paris, and Khaled Boussetta from Université Sorbonne Paris Nord. A special thank you to En Shao from the Institute of Computing Technology (ICT), China. And of course, we should not forget the extremely constructive influence of Kemal Ebcioglu.

A word of gratitude to the staff at the IEEE Computer Society and at IFIP for their indefatigable work. They immensely contributed to improving the scientific impact of NPC. We thank all the people we interacted with to improve the scientific impact of NPC. They will recognize themselves.

This conference was organized in cooperation with two IEEE Technical Committees (TCS) and two IEEE Special Technical Communities (STC):

- Technical Committee on Parallel Processing - TCPP, Anne Benoit, France.
- Technical Committee on Cloud Computing - TCCLD, Robert Hsu, Taiwan.
- Special Technical Community on Multicore - STC Multicore, Lawrence Rauchwerger, USA.
- Special Technical Community on Dataflow - STC Dataflow, Guang R. Gao, USA and Stéphane Zuckerman, France.

Finally, we would like to thank our home institutions for their support: CY Cergy Paris Université, Université Sorbonne Paris Nord (USPN) including the MathSTIC research group, ESIEE, ESME Sudria, and Université Gustave Eiffel. Thank you very much, we greatly appreciated their help and assistance.

A total of 62 submissions were received in response to our call for papers. These papers originated from Asia (China and Japan), Africa, Europe, and North America (USA). Each submission was sent to at least three reviewers, with an average of four reviewers per submission, and up to six reviewers. Each paper was judged according to its originality, innovation, readability, and relevance to the expected audience. Based on the reviews received, 20 full papers (about 32%) were selected to be published as LNCS proceedings. Among these, six papers were further selected to be extended and submitted to a Special Issue in the International Journal of Parallel Programming. Four main themes were identified during the conference: Networks and Communications; Storage; System

Software; and Applications and Algorithms. Two "Best Papers" sessions were also held, the first of which was oriented toward storage issues, while the other one was geared toward applications and algorithms.

We would also like to recognize our three guest speakers who gave exciting presentations: Yutang Lu, Sun Yat-sen University, China, who demonstrated current and future challenges to build and program supercomputers; Avi Mendelson, Technion, Israel, who spoke about data-centric computations for future high-performance systems; and Anne Benoit, ENS Lyon, France, who tackled the issue of resilience and fault-tolerance in current and future supercomputing systems.

Enjoy NPC 2021!

November 2021

Christophe Cérin
Depei Qian
Jean-Luc Gaudiot
Guangming Tan
Stéphane Zuckerman

Organization

NPC 2021 was jointly organized by ESME Sudria, ESIEE Paris, Université Gustave Eiffel, Laboratoire d'Informatique Gaspard Monge (LIGM), CY Cergy Paris Université, Université Sorbonne Paris Nord, and DNAC.

General Co-chairs

Christophe Cérin   Université Sorbonne Paris Nord, France
Depei Qian   Beihang University, China

Program Co-chairs

Jean-Luc Gaudiot   University of California, Irvine, USA
Stéphane Zuckerman   CY Cergy Paris Université, France
Guangming Tan   ICT, China

Local Arrangements Chair

Abdulhalim Dandoush   ESME Sudria, France

Publicity Co-chairs

Khaled Boussetta   Université Sorbonne Paris Nord, France
Chen Liu   Clarkson, USA
En Shao   ICT, China

Publication Chair

Éric Renault   ESIEE Paris, France

Web Chair

Elia Kallas   DNAC, France

Advisory Committee

Hai Jin (Chair)   Huazhong University of Science and Technology, China
Wenguang Chen   Tsinghua University, China

Yunquan Zhang	ICT, China
Weisong Shi	Wayne State University, USA
Shengzhong Feng	National Supercomputing Center in Shenzhen, China
Victor Prasanna	University of Southern California, USA

Steering Committee

Kemal Ebcioglu (Chair)	Global Supercomputing, USA
Hai Jin (Vice Chair)	Huazhong University of Science and Technology, China
Chen Ding	University of Rochester, USA
Jack Dongarra	University of Tennessee, USA
Guangrong Gao	University of Delaware, USA
Jean-Luc Gaudiot	University of California, Irvine, USA
Tony Hey	Science and Technology Facilities Council, UK
Guojie Li	ICT, China
Yoichi Muraoka	Waseda University, Japan
Viktor Prasanna	University of Southern California, USA
Daniel Reed	University of Utah, USA
Weisong Shi	Wayne State University, USA
Ninghui Sun	ICT, China
Zhiwei Xu	ICT, China

Technical Program Committee

Stéphane Zuckerman	CY Cergy Paris Université, France
Christophe Cérin	Université Sorbonne Paris Nord, France
Anna Kobusinska	Poznan University of Technology, Poland
Dezun Dong	National University of Defense Technology, China
Avi Mendelson	Technion, Israel
En Shao	ICT, China
Bo Yu	PerceptIn, China
Jean-Luc Gaudiot	University of California, Irvine, USA
Quan Chen	Shanghai Jiao Tong University, China
Éric Renault	ESIEE Paris, France
Philippe Clauss	Inria, ICube, University of Strasbourg, France
Bruno Raffin	Inria and University of Grenoble, France
Theo Ungerer	University of Augsburg, Germany
Weile Jia	ICT, China
Keqiu Li	Dalian University of Technology, China

William Jalby	Université de Versailles-Saint-Quentin-en-Yvelines, France
Weifeng Liu	China University of Petroleum, China
Hai Jin	Huazhong University of Science and Technology, China
Won Woo Ro	Yonsei University, South Korea
Claude Tadonki	Mines ParisTech, France
Jean-Thomas Acquaviva	DDN, France
Roberto Giorgi	University of Siena, Italy
Gabriel Paillard	Federal University of Ceará, Brazil
William Chu	Tunghai University, Taiwan
Shaoshan Liu	PerceptIn, China
Piyush Sao	Georgia Institute of Technology, USA
Heithem Abbes	University of Tunis El Manar, Tunisia
Mostapha Zbakh	ENSIAS, Mohammed V University of Rabat, Morocco
Keiji Kimura	Waseda University, Japan
R. Govindarajan	Indian Institute of Science, India
Albert Cohen	Inria, France
Sven Groppe	University of Lubeck, Germany
Felix Freitag	Universitat Politècnica de Catalunya, Spain
Marc Perache	CEA DAM, France
Arnaud Lallouet	Huawei Technologies Ltd, France
Cezary Mazurek	Poznan Supercomputing and Networking Center, Poland
Arthur Stoutchinin	ST Microelectronics, France
Pierre Manneback	University of Mons, Belgium
Patrice Darmon	Umanis R&D, France
Yiannis Papadopoulos	AMD, USA
Alba Cristina M. A. Melo	University of Brasilia, Brazil
Roberto Hsu	Asia University, Taiwan

Sponsoring Institutions

FR3734 MathSTIC USPN/CNRS

Contents

Algorithms and Applications

High Resolution of City-Level Climate Simulation by GPU with Multi-physical Phenomena

Koei Watanabe[1], Kohei Kikuchi[1], Taisuke Boku[1,2(✉)], Takuto Sato[2],
and Hiroyuki Kusaka[2]

[1] Master's Program in Computer Science, Graduate School of Systems
and Information Engineering, University of Tsukuba, Tsukubashi, Japan
{Kwatanabe,kikuchi}@hpcs.cs.tsukuba.ac.jp
[2] Center for Computational Sciences, University of Tsukuba, Tsukubashi, Japan
taisuke@ccs.tsukuba.ac.jp,
{sato.takuto.gu,kusaka.hiroyuki.ft}@u.tsukuba.ac.jp

Abstract. In this paper, we describe the Graphics Processing Unit (GPU) implementation of our City-LES code on detailed large eddy simulations, including the multi-physical phenomena on fluid dynamics, heat absorption and reflection by surface and building materials, cloud effects, and even sunlight effect. Because a detailed simulation involving these phenomena is required for analyses at the street level and several meters of resolution, the computation amount is enormous, and ordinary CPU computation cannot provide sufficient performance. Therefore, we implemented the entire code on GPU clusters with large-scale computing. We applied OpenACC coding to incrementally implement relatively easy programming and eliminate data transfers between the CPU and GPU memories. Based on this research, we determined that the elimination of data transfers is effective, even in the case where a part of the code execution on the GPU is slower than the CPU, owing to the absence of spatial parallelism. The objective of this study is to perform a complete climate simulation on a few square-kilometers field around the Tokyo Station, considering the finest resolution of the original highlighted area of the Marathon race in the Olympic Games Tokyo 2020. We successfully transferred the entire code to the GPU to provide approximately eight times the performance of CPU-only computation on multi-GPU per node with a large scale cluster.

Keywords: GPU · Large eddy simulation · Multi-physics · Climate · GPU cluster · OpenACC

1 Introduction

Graphics Processing Unit (GPU) computing is a prevalent method currently employed to achieve high-performance/power solutions, especially for large-scale cluster computing. GPU is a representative accelerating device for High Performance Computing (HPC), and exhibits a high degree of the Single Instruction Multiple Data (SIMD) parallelism with thousands of small cores and a high-memory bandwidth with the latest

C. Cérin et al. (Eds.): NPC 2021, LNCS 13152, pp. 3–15, 2022.
https://doi.org/10.1007/978-3-030-93571-9_1

memory technology, HBM2. Currently, the world's fastest supercomputer with GPU, Summit at the Oak Ridge National Laboratory, employs 4608 nodes, where each node is equipped with 6 NVIDIA Tesla V100 GPUs. Climate simulation with regular spatial grids is a representative application suitable for General Purpose GPU (GPGPU), and most of its calculations are performed in an SIMD approach (or SIMT on NVIDIA terminology). It is also appropriate to utilize a high-bandwidth memory (HBM2) under coalesced access, via regular vector processing. Moreover, recent GPUs have been equipped with large amounts of memory, e.g., up to 80 GB per GPU, to facilitate large-scale simulation.

Although GPU computing robustly supports the regular parallelization with high bandwidth memory, several issues adversely affect the actual porting of GPU codes from the original CPU codes. In addition, the data copy between the CPU and GPU memories poses a significant challenge. Because their address spaces are separated, we need to synchronize the data between them when GPU kernels are partially implemented. In OpenACC2.0 coding, the data image between CPU and GPU memories can be virtually shared by the BIOS support; however, the data are physically transferred under the surface, which triggers performance degradation; and the user is unconscious with this phenomenon.

Because the computation performance of a GPU's single core is significantly lower than that of a high-end CPU, GPU is typically not applied for a small amount of computation, especially for partial sequential execution in a large code. However, we need to consider such a small computation, which even triggers a substantial overhead of data transfer between the CPU and GPU. Accordingly, we need to examine the data transfer cost to be included in the total cost of computation. In addition, our target code named City-LES [1] handles the multi-physical phenomena for the precise analyses of temperature and wind flow in an urban area, such that the simulation complexity is significantly higher than that of a conventional climate simulation with fluid dynamics. In this study, we focused on the data transfer cost, as well as the computation time, and inferred that the complete porting of the entire code to the GPU is the best approach for such a multi-physical climate simulation. We implemented and evaluated the performance of GPU-ready City-LES and applied it to actual and practical simulations in the Tokyo Station area.

The contribution of this study is as follows:

(1) Elucidated the better performance of complete GPU-ported coding, even with several low-performance parts on the GPU, than that of the CPU when the overhead data transfer between the CPU and GPU is significant,
(2) Validated the performance improvement achieved by GPU for both the simple fluid dynamics code and complicated climate simulation with multi-physical phenomena, and
(3) Performed a real-world detailed climate simulation with multi-physics via the fully GPU-ported climate simulation, which is applicable in any other urban area simulation toward city modeling against the high temperature in human life.

2   City-LES Simulation Code

City-LES [1] is a large-eddy simulation (LES) model that simulates the microscale weather and climate in an urban district with extremely high spatial resolutions (approximately 1–5 m). This model is a hybrid of meteorological and computational fluid dynamics (CFD) models. Its specific features include: (1) resolving buildings explicitly, (2) considering the thermal and mechanical effects of street trees, (3) explicitly representing shading by buildings and street trees, (4) considering multiple reflections of radiation by buildings and trees, and (5) incorporating cloud microphysics and atmospheric radiation models. These features facilitate the evaluation of thermal and wind environments, especially in urban districts, considering the effects of roadside trees and building shadows with high spatial resolutions. In addition, City-LES can quantitatively evaluate the effect of urban heat island mitigation strategies, such as dry-mist. These features are valuable for scientists and policy decision-makers.

To simulate the thermal environment in an urban area using LES, we need a model that can calculate the three-dimensional radiative processes between buildings and roads. City-LES combines a radiative model using the radiosity method as a model for calculating such complex radiative processes. The radiosity method can calculate the radiance of even complex city blocks with high accuracy. Surface temperature, as well as sensible and latent heat fluxes, can be calculated based on the calculated values of shortwave and longwave radiation, and when combined with the LES model, the detailed thermal environment of an urban district can be reproduced with a spatial resolution of approximately 1–5 m.

Originally, we implemented the basic code extension by GPU from the CPU code which solely adopts OpenMP + MPI [2]. In this implementation, several features, such as surface radiation tracking, are simplified and remain on the CPU, while other functions run on the GPU. When we apply the complete and practical model of an actual urban area, such as Tokyo City, data transfer between CPU and GPU occurs in a "reverse-offloading" manner. In this study, we implement a complete set of City-LES functions on a GPU to avoid data movement and apply the code to an actual urban model.

The temperature on the ground surface, especially in the deep summer season, significantly affects human life, and a severe social challenge is posed by global and local heating. In this study, we implemented a complete code by GPUs and evaluated its performance using a model of the Tokyo Station area, which was originally planned to be the objective of the Marathon race in the Olympic Games Tokyo 2020. Unfortunately, the game was postponed by one year, and the venue was changed to Sapporo City in Hokkaido Island; however, we remain interested in the climate conditions of this representative space in the gigantic city of Tokyo. Primarily, the venue was changed because of the estimation of extremely high temperatures in Tokyo, which is an optimal target for our City-LES application. Figure 1 illustrates an example of the image results around the Tokyo Station, which was created by the CPU version of City-LES.

Fig. 1. Simulation result image of City-LES

The original version of the City-LES code was developed for general-purpose CPUs with parallelization, using OpenMP + MPI in FORTRAN. In most cases, the basic construction of the code comprised three-level nested loops for the 3D structure of the space domain, considering the surrounding area of data in 3D arrays, similar to conventional 3D stencil calculations; therefore, 3D neighboring communication on the MPI was required with blocked domain decomposition. Within each MPI process, most of the parts were parallelized at the thread level using OpenMP. The 3D domain was parallelized in the 2D X- and Y-dimensions, whereas the Z-dimension was not parallelized because the grid size of the Z-dimension was usually smaller than that of the X- or Y-dimension, and we need to keep a moderate size of process grain without too much number of processes. Hence, the main communication between MPI processes emerged as the 2D nearest neighboring communication.

3   GPU Coding and Optimization

3.1   Basic GPU Coding for Kernels to Implement Functions in City-LES

Hereafter, we adopt the terms "GPUize" and "GPUization" for code porting from CPU to GPU. We applied OpenACC + MPI programming for GPUization from the original OpenMP + MPI code. In this ssssection, we briefly describe our previous work on basic GPUization [2], as it contains necessary information for understanding this paper. Basically, we applied the OpenACC directive instead of OpenMP in the original CPU code to handle a single GPU per MPI process. Accordingly, the multithread execution on a multicore CPU corresponded to multicore execution on a GPU. We have not changed any data structure and loop construction from the original CPU version, and just have modified OpenMP directives by OpenACC ones with supplementary directives and clauses.

Figure 2 presents the basic structure of the function modules in City-LES, with a time breakdown on the original CPU code on an Intel Xeon E5–2690 (2.6 GHz) with

14 cores × 2 sockets, where we ran it with 14 threads (OpenMP) × 2 processes (MPI) on a relatively small problem size of 128 × 128 × 128 (X × Y × Z). The breakdown was sorted in the order of the time consumption of each part. Here, we introduce the major time consuming functions.

- *MGOrthomin_m*
  solves Poisson Equation by multigrid Orthomin(1) method [5] with pre-conditioning, by stencil computation and nearest neighboring communication for sleeve zone.
- *sgs_driver*
  calculates the subgrid scale turbulence, with mostly the matrix summation and stencil computation.
- *update_rk_u*
  calculates the approximate result by Runge-Kutta method, with the calculation of average of matrix, allreduce communication and stencil computation.
- *check*
  checks the computation normality by CFL condition, which max/min calculation of the matrix and allreduce communication.
- *advection*
  calculates the advection term, which stencil computation.

The parallelization and GPUization of *MGOrthomin_m* is a difficult task, as it is not a simple 3D nested loop. All other functions with red-colored names in the legend can be easily GPUized by simply applying the computation grid mapping of the GPU to the appropriate level of the nested loop. To increase the parallelism and utilize several thousands of cores for the GPU, we expanded the nested loop into a single dimension via loop coalescence when necessary.

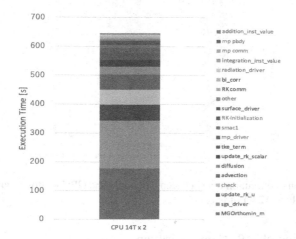

Fig. 2. Breakdown of execution time in City-LES functions (14 threads × 2 processes on two CPUs).

After the GPUization of the other functions in the 3D nested loop, where a simple OpenACC kernel description works optimally, several small functions remained in the CPU owing to the insufficient performance of the GPUization triggered by the absence of parallelism. However, we determined that it is significantly better to port all these functions in GPU kernels, even if the execution time of the function is longer than that of the CPU version to reduce the total execution time by eliminating the data transfer between the CPU and GPU memories. It is an advantage of OpenACC rather than that of CUDA on NVIDIA GPUs, where we can easily describe any part, including a sequential execution part, as a GPU kernel. It is usually not applied to the inefficiently sped up part of the GPU; however, we have the following policy:

> *Although the execution time for each minor function on the GPU is longer than that of the CPU version, to shorten the total execution time, it is better to prevent the data copy between CPU and GPU , and OpenACC is significantly easier than CUDA coding.*

Figure 3 presents the comparison of execution time for 200 time steps on a small problem size with CPU-only, GPUization with data movement, and GPUization without data movement. In the case without data movement, all the functions in the main execution part were implemented on the GPU, except the initialization and file I/O parts, which remained on the CPU with a negligible fraction. In the case with data movement, the green and sky-blue colored bars (top two) represent the data transfer overhead between the CPU and GPU. In the code with complete GPUization (right), the execution time reduced to 35% of the partially GPUized version (middle). Finally, the execution time reduced to 16% of the original CPU-only version (left). Then, we ported the remaining minor functions as small kernels using OpenACC, and coded the completely GPUized version of City-LES, which we call City-LES/GPU.

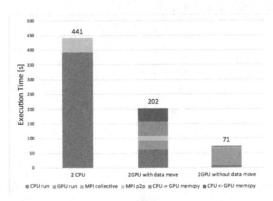

Fig. 3. Comparison of execution time between CPU-only and GPU with/without data movement on small size problem (128^3) of execution

3.2   Optimization of MPI Communication

Because our MPI level parallelization is performed on 2D blocking (as described in the last part of Sect. 2), the nearest-neighbor communication on the surface of the 3D region indicates the block-stride and stride communication patterns. To minimize the communication cost, we applied hand-written packing and the unpacking of non-contiguous data on GPU memory before and after the communication. It is well known that the type-vector feature of MPI communication is ineffective on GPUs, and we verified that our proposed method is faster than the type-vector implementation.

For the latency hiding of point-to-point MPI communication between neighboring nodes for stencil computation, we applied asynchronous communication, where the data on edge parts were initially transferred, before they reached the destination, while inner points were computed without communication. We also applied GDR (GPU-Direct RDMA) [8] for a short latency of communication between GPUs over nodes.

4 High Resolution of Detailed Simulation on Actual City

In this research, we evaluated the performance of City-LES/GPU for practical use to target the Tokyo Station area in central Tokyo City, which was originally planned as the highlighted zone of the Marathon race in the Olympic Games TOKYO 2020. Because the August temperature (approximately 35 °C) and humidity (more than 80%) of Tokyo City were extremely high, it was important to estimate the temperature distribution of that area, to consider the health of a large number of audiences watching the game. Unfortunately, the game was postponed to 2021, and eventually, the venue was moved to Sapporo City on the northern island (Hokkaido) to achieve the intended goal. However, we still believe that this simulation is important because we already know that the temperature distribution varies at each point, owing to the complicated structure of buildings, streets, and trees in that area. In addition, because City-LES can create wind flow along the streets, as well as temperature, the analysis can be adopted to remodel the city's construction. Finally, an objective of this study is to apply City-LES/GPU to heat problems in big cities, for future urban structure design and solutions.

As described in Sect. 2, City-LES can handle detailed city construction, including the surface material, unlike conventional climate simulation via fluid dynamics; in addition, it can calculate the sunlight's movement, which significantly affects the temperature transition. We completely GPUized all the functions for general GPU clusters using OpenACC + MPI. To perform the 5-m grid size simulation of the Tokyo Station area, we created a $512 \times 512 \times 512$ grid size model based on a real 3D map of that area. This is the same model illustrated in Fig. 1 where the image was created by the original City-LES (CPU version).

In the follow-up study after [2], we modified another major function named *surface_driver*, which adopts the radiosity method and surface material handling for the reflection and absorption of sunlight to calculate the heating effect by sunlight. This part is also complicated because it refers to the surface material on each grid point that is covered by a concrete building block, grass, or tree. Here, the building blocks and trees have a 3D structure, and the continuous 3D grid points of the same material are reconstructed as a single object in the inner data format.

5   Performance Evaluation

5.1   Target Supercomputer

In this section, we evaluate the City-LES/GPU code on an actual supercomputer with a multi-GPU configuration in each node. We applied our code to the multi-hybrid accelerated supercomputer Cygnus [8] at the Center for Computational Sciences at the University of Tsukuba. We employed up to 64 Cygnus nodes, out of a total of 81 nodes. Table 1 presents the basic specifications of Cygnus. Each node is equipped with two CPU sockets, four GPU cards, and two optional FPGA cards. In this research, we adopted GPUs as accelerators and did not use FPGA parts. Table 1 presents the specifications of the Cygnus cluster. Although the 32 nodes of Cygnus consist of two FPGAs (Intel Stratix10), we did not use them for this research.

City-LES/GPU was coded to adopt an equal number of CPU cores and a single GPU for each MPI process. Because a single node of Cygnus comprises 24 cores (12 cores × 2 sockets) of CPU and 4 GPUs, we ran four MPI processes on each node, with six cores (threads) of CPU execution and a single NVIDIA Tesla V100 GPU. To examine the strong scaling in parallelization, we scaled the system size from 4 to 64 nodes and 16 to 256 GPUs on a single size problem with $512 \times 512 \times 512$ grids. All computations were performed with double precision (FP64). The space resolution of the model was 5 m × 5 m × 5 m, and the time resolution (Δt) was 0.1 s. We performed 600 time steps (60 s) of the simulation on the model. We evaluated the results of City-LES/GPU, compared it with the CPU version, and verified that the result fits it with a negligible fraction of error, at a maximum of approximately 0.2 °C. We believe that such an error is triggered by the computation order change in the GPU execution.

Table 1. Cygnus cluster specifications

CPU/node	Intel Xeon Gold 6126 (12 cores) × 2
CPU memory/node	DDR4–2666 192 GB (16 GB × 6 chan. × 2 sockets)
GPU/node	NVIDIA Tesla V100 (PCIe, 32 GB) × 4
Interconnect	InfiniBand HDR100 × 4
# of nodes	81
Theoretical Peak Perf	2.43 PFLOPS (double precision)
Host OS	CentOS 7.9
Compiler	Nvidia HPC SDK 21.2
MPI library	OpenMPI 3.1.6
CUDA version	11.2

5.2   Overall Performance

Table 2 presents the general execution time of City-LES and City-LES/GPU for the target problem, scaling the number of nodes from 4 to 64 for strong scaling. In this paper, we focus on the complete elimination of data movement between the CPU and GPU, therefore we compared the last version of City-LES/GPU where *surface_driver* and *advection* functions are remained on the CPU and full GPU version presented in this paper. Hereafter, the former version is called CL/CPU while the latter is called CL/GPU. Based on the performance of the 4-node case, the parallel efficiency of CL/CPU varies from 1.22 to 1.99, while that of CL/GPU varies 0.94 from 2.59. In both CL/CPU and CL/GPU, the parallel efficiency is optimal, especially for CL/GPU. This is because the performance of the 4-node case on the CL/GPU is relatively low. Although a more detailed analysis is currently being conducted, we hypothesize that this phenomenon is primarily caused by the lack of registers and low cache hit ratio triggered by a large grid size per GPU. However, it can be observed that the scalability of City-LES by both CPU and GPU is remarkable, even with strong scaling.

Figure 4 shows the execution time and performance gain for CL/GPU over CL/CPU. The maximum performance gain is achieved with the 8-node case, where CL/GPU achieves 7.9 times the performance of CL/CPU. This indicates that our CL/GPU is substantially effective, even for a complicated city model where the computation efficiency decreases in several functions (refer to Sect. 5.3). The performance gain of more than 8 nodes was reduced to 64 nodes; however, the scalability of CL/GPU still maintained a high-parallel efficiency of 94%.

Table 2. Performance of City-LES by CPU and GPU

#nodes	Exec. Time (sec)		Spped-up (to 4 nodes)		Parallel efficiency	
	CL/CPU	CL/GPU	CL/CPU	CL/GPU	CL/CPU	CL/GPU
4	7226.7	1954.6	1.00	1.00	1.00	1.00
8	2966.9	376.8	2.44	5.19	1.22	2.59
16	1260.2	201.4	5.73	9.71	1.43	2.43
32	454.1	141.1	15.91	13.85	1.99	1.73
64	252.3	130.4	28.64	14.98	1.79	0.94

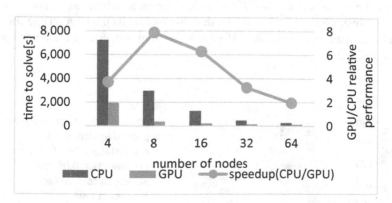

Fig. 4. Overall execution time of CL/CPU and CL/GPU with relative performance gain of GPU over CPU

5.3   Performance Bottleneck on City-LES/GPU Scalability

As described in Sect. 5.2, we observed that the performance scalability on strong scaling is weaker in CL/GPU than in CL/CPU. This can be explained by the fact that the computation performance on the GPU is significantly higher than that of the CPU, which makes the communication overhead on MPI visible on the CL/GPU. Figure 5 shows the breakdown of the execution time with 64 nodes, where the performance difference between CL/CPU and CL/GPU is the smallest. It can be observed that most of the functions were accelerated via GPU computation. In particular, the performance gain in the *surface_driver* (colored in orange) traces the sunlight movement every 5 min (simulation time) to calculate the heat effect on the grids, according to their materials. In fact, although the *surface_driver* is substantially expensive, owing to heavy computation, we solely implemented this function once in every 3000 time steps (5 min of simulation time when $\Delta t = 0.1$ s) because the sunlight movement is sufficiently small in that duration.

This portion comprises a radiosity process that is highly effective in GPU computing. Most of the other major functions are the scanning and computation of 3D grid data, where the computation power and high memory bandwidth obtained by HBM2 are robust to achieve an optimal performance gain.

In contrast, the *Smac1* function (colored in sky-blue) for calculating the Poisson equation on the grids exhibits no performance gain on CL/GPU. We measured the execution time of this function for various numbers of nodes and deduced that the performance gain was also negligible. To analyze the root cause, we examined the computation and communication times for each case. Figure 6 illustrates the breakdown of the *MGOrthomin* subfunction for the Poisson solver, which dominates the execution time in *Smac1*. For this Poisson solver, we applied the multi-grid method presented in [5]. In the "V-style" multi-grid method, the computation amount reduced half by half in multiple sub-steps, and each sub-step required a certain size of data communication among MPI processes. As illustrated in Fig. 6, the communication time dominates the execution of the *MGOrthomin* function, where we cannot scale the performance over 16 nodes. We consider that this part limits the CL/GPU performance, while the other parts are accelerated. Hence, reducing this overhead is a crucial priority for future research.

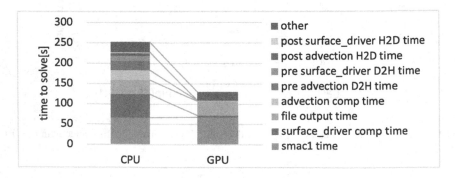

Fig. 5. Breakdown of execution time

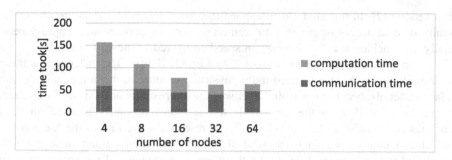

Fig. 6. Computation and communication times on the Poisson solver in CL/GPU (*MGOrthomin*)

In Fig. 5, another large fraction of the execution time is occupied by the *file output* (depicted in gray). This function outputs the simulation results at certain time steps as a combined structured file for each MPI process. We can hide this cost because the function is solely for the file output, thus we estimate that the execution time can be reduced by approximately 35% from this graph in our next work.

Another factor affecting the performance variation is the complexity of the simulation target model. For a small problem size of 128^3, we examined the execution time for three models: single building in the space, equally distributed number of buildings in the space, and equally distributed number of buildings with adjacent trees. The execution time increased in the following order: 43.3 s, 54.6 s, and 57.4 s, where the second and third models increased the execution time to that of the first one at + 26% and + 33%, respectively. This increase was triggered by the processing of the *Smac1* function for the Poisson solver, owing to the longer convergence time. The obtained results indicate that our modified Poisson solver is applicable for a complex city model that negligibly increases the execution time.

6 Related Works

Researchers at the Tokyo Institute of Technology implemented the entire code of the practical climate simulation code that was developed at the Japan Meteorological Agency, which is called ASUCA by CUDA [3]. In this research, the complete version of the acting climate prediction code was ported to the Tsubame 2.0 supercomputer at Tokyo Tech. Owing to the unavailability of a CUDA FORTRAN compiler, the authors converted the original FORTRAN code to the CUDA C code. They employed 3936 GPUs (Fermi GPU), utilizing 76.1 TFLOPS of the sustained single-precision performance.

In a study conducted by Onodera et al., the LES code was developed to simulate the 10×10-km region of urban Tokyo, with a 1-m mesh size [4]. On the Tsubame 2.0 system, they achieved 198 GFLOPS on each GPU with single precision, up to 149 TFLOPS of sustained performance with 1000 GPUs.

RIKEN R-CCS developed a large-scale climate simulation program called Scalable Computing for Advanced Library and Environment (SCALE) and implemented it on a

GPU version [7]. In this study, they applied the computation reordering to utilize the locality of data access or connect the kernels to achieve performance improvement. Finally, they achieved a 5 × increase in speed compared to the CPU version.

The primary difference between our City-LES/GPU from the programs of these previous studies is that we inferred multi-physical phenomena, including sunlight and surface materials, because our main target was the temperature analysis of the surface, as well as wind flow in the city, while these studies primarily focused on fluid dynamics computations. As a novel study, our research is the first in the literature to port such a complicated multi-physical simulation of urban climate with a GPU-accelerated supercomputer. In addition, these previous studies did not specify how to reduce the data copy between the CPU and GPU memories, as well as the concept of introducing OpenACC completely; furthermore, they did not specify how to decelerate the partial code and achieve a total performance improvement, which indicates the novelty of this present research.

7 Conclusions

In this study, we implemented the GPUized version of the city-level climate code City-LES, developed at the Center for Computational Sciences at the University of Tsukuba. The GPU version of the City-LES/GPU code was completely accelerated by OpenACC for large-scale GPU clusters, and we applied it to an actual city model of the Tokyo Station area with 512 × 512 × 512 3D grids for a 5-m mesh size and high-resolution simulations on a supercomputer (Cygnus), scaling up to 64 nodes with 256 T V100 GPUs. To obtain an adequately detailed simulation of the wind flow and temperature of a city, City-LES and City-LES/GPU were designed to cover the surface materials on the ground, radiation of sunlight, and building blocks. The obtained simulation results exhibited significantly high scalability with strong scaling up to 256 GPUs. With this City-LES/GPU, we can execute a 60 s simulation case with only 130 s in the actual computation time. This result substantially improves climate research.

In the future, our research will include the optimization of the communication time in the Poisson solver, which limits the performance scalability, apply City-LES/GPU to various urban simulations, and utilize this tool for future urban design to improve human life.

Acknowledgment. The authors thank our colleagues for their substantial support and suggestions, especially Mr. Ryosaku Ikeda at Weathernews Inc., Prof. Doan Quang Van, Prof. Norihisa Fujita, and Prof. Ryohei Kobayashi, all in the CCS at the University of Tsukuba. This research was partially supported by the MEXT "Next Generation Supercomputing" program, with the titled project "Development of Computing-Communication Unified Supercomputer in Next Generation." In addition, this study was supported by MCRP at CCS, University of Tsukuba, which provided the Cygnus supercomputer used in the research, under the "HPC application and system software development on FPGA-GPU combined platform" Project.

References

1. Ikeda, R., Kusaka, H., Iizuka, S., Boku, T.: Development of urban meteorological LES model for thermal environment at city scale. In: Proceeding of the 9th International Conference for Urban Climate, Toulouse, France (2015)
2. Tsuji, D., Boku, T., Ikeda, R., Sato, T., Tadano, H., Kusaka, H.: Parallelized GPU code of city-level large eddy simulation. In: Proceeding of International Symposium on Parallel and Distributed Computing (ISPDC), Warsaw, Jun. 2020 (2020)
3. Aoki, T., Shimokawabe, T.: Chapter 16: Of GPU solutions to multi-scale problems in science and engineering: large-scale numerical weather prediction on GPU supercomputer. In: David, A.Y., Wang, L., Chi, X., Johnsonn, L., Ge, W., Shi, Y. (eds.) Lecture Notes in Earth System Sciences. Springer, Berlin (2013)
4. Ahmad, N.H., Inagaki, A., Kanda, M., Onodera, N., Aoki, T.: Large eddy simulation of the gust factor using Lattice Boltzmann method within a huge and high resolution Urban Area of Tokyo. J. Jpn. Soc. Civil Eng. Ser. B1 71(4), I_37–I_42. https://doi.org/10.2208/jscejhe.71.I_37
5. Tadano, H., Ikeda, R., Kusaka, H.: Speeding up Large Eddy Simulation by Multigrid Preconditioned Krylov Subspace Methods with Mixed Precision. In: The 35th JSST Annual Conference International Conference on Simulation Technology (JSST2016), Kyoto, Japan (Oct. 2016)
6. Nvidia: CORPORATION: PGI, Resources, Cuda Fortran. https://www.pgroup.com/resources/cudafortran.htm
7. Team-SCALE R-CCS RIKEN, SCALE by Riken R-CCS. http://r-ccs-climate.riken.jp/scale/a/index.html
8. Cygnus Supercomputer. https://www.ccs.tsukuba.ac.jp/wp-content/uploads/sites/14/2018/12/About-Cygnus.pdf

dgQuEST: Accelerating Large Scale Quantum Circuit Simulation through Hybrid CPU-GPU Memory Hierarchies

Tianyu Feng[1,2], Siyan Chen[2], Xin You[2], Shuzhang Zhong[2],
Hailong Yang[1,2(✉)], Zhongzhi Luan[2], and Depei Qian[2]

[1] State Key Laboratory of Software Development Environment, Beihang University,
Beijing, China
hailong.yang@buaa.edu.cn
[2] School of Computer Science and Engineering, Beihang University, Beijing, China

Abstract. With the advancement of quantum computing, verifying the correctness of the quantum circuits becomes critical while developing new quantum algorithms. Constrained by the obstacles of building practical quantum computers, quantum circuit simulation has become a feasible approach to develop and verify quantum algorithms. Although there are many quantum simulators available, they either achieve low performance on CPUs, or limited simulation scale (e.g., number of qubits) on GPUs due to limited memory capacity. Therefore, we propose *dgQuEST*, a novel acceleration method that utilizes hybrid CPU-GPU memory hierarchies for large-scale quantum circuit simulation across multiple nodes. *dgQuEST* adopts efficient memory management and communication schemes to leverage the distributed CPU and GPU memories for accelerating large-scale quantum simulation. Our evaluation demonstrates that *dgQuEST* achieves an average speedup of 403× compared to *QuEST* on quantum circuit simulation with 32 qubits, and scales to quantum circuit simulation with 35 qubits on two GPU nodes, far beyond the state-of-the-art implementation *HyQuas* can support.

Keywords: Quantum simulation · Distributed GPU acceleration · Memory and communication optimization

1 Introduction

With the development of quantum algorithms, quantum computing has become the most likely means to surpass the performance of traditional computing in the future. However, it is difficult to verify quantum algorithms due to the still early age of building practical quantum computers. Therefore, using traditional computers to simulate the operations of quantum circuits has become one feasible approach to develop and verify new quantum algorithms. In general, quantum simulation methods are mainly divided into two categories, such as state vector

© IFIP International Federation for Information Processing 2022
Published by Springer Nature Switzerland AG 2022
C. Cérin et al. (Eds.): NPC 2021, LNCS 13152, pp. 16–27, 2022.
https://doi.org/10.1007/978-3-030-93571-9_2

(full amplitude) method and tensor network method. The scalability of the state vector method is limited by the number of qubits, as each additional qubit will double the memory usage. The tensor network method uses tensor shrinkage and decomposition to avoid the exponential memory bloating, however it suffers poor performance for deep circuits or circuits with higher entanglement. Many quantum circuit simulation software have been developed over years, including *Qibo* [3], *QuEST* [7], *qHipster* [10], *AC-QDP* [6], *QuiMB* [5], *TN QVM* [9], etc. Among them, *QuEST* [7] is one of the most widely used high-performance quantum circuit simulators, and a large amount of research work has been carried out based on *QuEST* [1,4,8,12].

QuEST [7] is designed based on the full-amplitude quantum simulation method and supports single-qubit gates with multiple control qubits. Among them, the application of the qubit gate is carried out by applying corresponding computations on the entire state vector in sequence. The computations on the state vector are paired up and each pair is independent from others. Therefore, the simulation of *QuEST* can be easily parallelized on GPU. However, the parallel simulation of *QuEST* can only be supported on a single GPU, which constrains the size of quantum circuit can be simulated due to the limited GPU memory. The latest improvement on *QuEST* such as *HyQuas* [12] can speedup the simulation on multiple GPUs. However, it fails to exploit the hybrid CPU-GPU memory hierarchies, and thus requires more GPUs to support large-scale quantum circuit simulation. Therefore, we propose *dgQuEST*, a novel acceleration method that exploits hybrid CPU-GPU memory hierarchies with efficient memory management and communication schemes to improve the performance of large-scale quantum circuit simulation.

Specifically, this paper makes the following contributions:

- We propose a CPU-GPU hybrid memory management scheme, which effectively utilizes the large capacity of CPU memory as well as the high performance of GPU memory for large-scale quantum simulation.
- We propose a page-table based memory management scheme to manage the qubit mapping of the entire state vector, which improves the data locality of the state vector access for better performance.
- We propose a pipelined communication scheme to reduce the overhead of distributed memory access, which further improves the performance of *dgQuEST* when scaling to multiple GPU nodes.

2   Background and Motivation

2.1   Full-amplitude Quantum Simulator

A full-amplitude quantum simulator (FAQS) stores all the amplitudes in a state vector SV and a quantum gate can be represented as a matrix transformation. Assume that the quantum register R has n qubits and the SV represents the current quantum state $|\Psi\rangle$, i.e. $|\Psi\rangle$ can be represented as in Eq. (1):

$$|\Psi\rangle = \sum SV_{q_{n-1}q_{n-2}\ldots q_1 q_0} |q_{n-1}q_{n-2}\cdots q_1 q_0\rangle \tag{1}$$

where $q_i \in \{0,1\}, i \in \{0,1,\ldots,n-1\}$. If a quantum NOT gate, which is the same as binary NOT operations in classical computers, is applied on the xth qubit (numbered from 0), the state vector SV will be transformed to SV' as shown in Eq. (2), where $q_i \in \{0,1\}, i \in \{0,1,\ldots,x-1,x+1,\ldots,n-1\}$.

$$\begin{pmatrix} SV'_{q_{n-1}q_{n-2}\cdots q_{x+1}0q_{x-1}\cdots q_1q_0} \\ SV'_{q_{n-1}q_{n-2}\cdots q_{x+1}1q_{x-1}\cdots q_1q_0} \end{pmatrix} = \begin{pmatrix} 0 & 1 \\ 1 & 0 \end{pmatrix} \begin{pmatrix} SV_{q_{n-1}q_{n-2}\cdots q_{x+1}0q_{x-1}\cdots q_1q_0} \\ SV_{q_{n-1}q_{n} \; {}_2\cdots q_{x+1}1q_{x-1}\cdots q_1q_0} \end{pmatrix} \quad (2)$$

Therefore, performing a single-qubit gate transformation introduces data dependencies between elements in the state vector in pairs. In detail, the distance of the interdependent elements in a pair is 2^i when the gate performs on the i-th qubit, which leads to poor spatial locality in memory. Besides, when a series of quantum gates perform on different qubits separately, the data-dependent structures changes completely during each computation, leading to higher cache miss within the existing memory hierarchy.

2.2 Optimizing FAQS on Distributed CPU Nodes

For quantum simulations that exceed the available memory, FAQS needs to distribute the simulation to fit in the memory available on each node. In detail, the state vector is split into blocks stored separately in each node and the entire state vector needs to be updated when applying a transformation of a gate to the current state. However, the interdependent data may be stored separately in different nodes. Figure 1(a) demonstrates an example of data dependence across nodes in a two-node distributed FAQS, where each node depends on all the data on the other node to finish the simulation of a quantum gate on the 3rd qubit. Figure 1(b) demonstrates a more general case when two single-qubit gates are performed on the higher two qubits, respectively. The *peer* relationships are established between nodes two by two based on the target qubit of the gate and each node needs to obtain all the data of the *peer* to compute its local result.

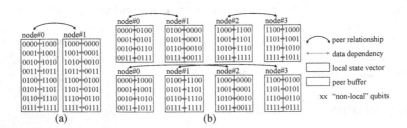

Fig. 1. Peer data interaction in a distributed implementation of FAQS. The simulated quantum register is assumed to contain 4 qubits and distributed to (a) two nodes, each storing 8 state vector elements (3 qubits), and (b) four nodes, each storing 4 state vector elements (2 qubits), with two single-qubit gates acting on the third and fourth qubits shown in the upper and lower halves of the graph, respectively.

2.3   Optimizing FAQS on GPUs

There are several research works to optimize FAQS on GPUs. *HyQuas* combines two existing GPU optimization methods such as *ShareMem* and *BatchMV*, both of which take multiple gates and calculate them together to reduce memory access. The *ShareMem* method loads interdependent data required by the quantum gates into shared memory as a local state vector and sequentially applies quantum gate transformations. Whereas the *BatchMV* method merges the gates into a larger matrix and applies a matrix-vector multiplication between the matrix and the local state vector. Utilizing high speed shared memory, these two methods greatly reduce memory access overhead. *HyQuas* further improves the above two methods and implements simulation on multiple GPUs with optimized data transfer mechanism. However, *HyQuas* requires to store all data into GPU global memory, which strictly limits the number of qubits for simulation if the number of nodes is limited. *Qibo* also supports accelerating quantum simulation on multiple GPUs, but achieves poor performance and high memory usage due to frequent data exchange between CPU and GPUs. On contrast, *dgQuEST* leverages the hybrid CPU-GPU memory hierarchies across multiple nodes efficiently, which accommodates accelerated large-scale quantum circuit simulation.

3   Methodology

3.1   Design Overview

According to the memory hierarchy, *dgQuEST* applies a *three-level memory model* to divide the distribution of state vectors into three top-down levels: CPU main memory, GPU global memory, and shared memory. Figure 2 demonstrates the overall design of *dgQuEST*, including *memory manager*, *gate aggregator*, *task dispatcher*, *executor*, *data reorganizer*, and *communicator*. The *memory manager* maintains the state vector in page tables. The *gate aggregator* is used to reorder the quantum gates when the storage hierarchy is lowered to reduce the number of interactions. Besides, *dgQuEST* also applies *three-level memory model* inside *memory manager* and *executor*, and utilizes *gate aggregator* for *circuit partitioning* along this *three-level memory model*. The *task dispatcher* aims to assign task blocks in main memory to different GPUs on each node. The *executor* performs the actual quantum gate operations with the assistance of *gate aggregator* to decide which data to load into shared memory. After computation, the data is reorganized with *data reorganizer* according to the target qubits of the current circuit and the next circuit to match the new dynamic mapping between qubits and state vector index bits. The reorganized data is sent to the corresponding node via *communicator* or written back into the current node, both with new page numbers calculated from the result of remapping.

3.2   Page-Table Based Memory Management

To achieve efficient data exchange between nodes, we implement page-table based memory management mechanisms and maintain a memory pool for pages.

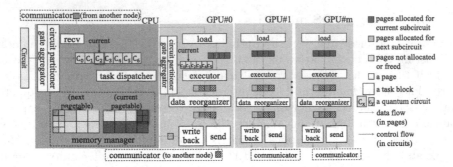

Fig. 2. Overall design of *dgQuEST*. For simplicity, data flows and control flows are omitted except for GPU#0.

Specifically, the structure of the entire m-bits state vector index is divided into *page number* (higher $2k$ bits) and *offset* (lower $m - 2k$ bits), where the number of *page number* bits is always even and is divided into two halves. The lower half and the page offset jointly form the GPU explicit memory index, whose value range is the memory block size of a GPU calculation task. The higher half is further divided into the task number for *task division*. In general, the page is the smallest unit of data migration. Using page tables has the following advantages: *1)* The page enables fine-grained buffering and scheduling for communication. During data exchanges, once the node successfully transfers a page in the main memory to the GPU global memory, the page can be released to receive new data. Thus, only a small number of caching pages are needed for data exchange, which effectively reduces the memory overhead for communication. *2)* The logical page mapping of the page table enables flexible dynamic qubit mapping for better data locality. Specially, when the qubit mapping change occurs only in the page number part of the state vector index, the dynamic mapping can be completed only by modifying the page table without any data exchanges. *3)* The page number provides sufficient information (task number and node index) of the data block for the communicator to confirm the destination and source of the data during data exchange.

3.3 DAG-Based Gate Aggregation

dgQuEST analyzes the dependencies between quantum gates in a quantum circuit to construct a directed acyclic graph (DAG). Then *dgQuEST* applies a heuristic algorithm to derive several aggregated sub-circuits from the DAG. In general, *gate aggregation* reorders the quantum circuit and splits it into as few sub-circuits as possible under the premise that the number of target qubits of each sub-circuit does not exceed a given threshold.

DAG Construction - Each node in the DAG corresponds to a quantum gate in the quantum circuit and each directed edge indicates a directed dependency between the quantum gates. We assume that the control qubit set of the quantum gate corresponding to node P is C_P and the target qubit set is

T_P. For all nodes A and B such that A appears earlier than B in the original execution order, there is a directed edge from A to B, iff there exists a qubit $x \in (C_A \cup T_A) \cap (C_B \cup T_B)$ such that for all gate C appears between A and B, $x \notin C_C \cup T_C$. Therefore, for all nodes in the constructed DAG, the quantum gate corresponding to one node must be calculated before all the quantum gates corresponding to its directed linked nodes. All computation reorderings that guarantee this principle will result in mathematically equivalent quantum circuit simulations. To construct a DAG, we maintain a vector storing heading gates on each qubit (the last gate whose target or control qubits contain that qubit). When adding a gate G, we add directed edges from the heading gates on qubits in $C_G \cup T_G$ to G and make G as the new heading gate on these qubits.

Greedy Sub-circuit Partitioning - According to the *three-level memory model*, a lower memory level has lower data access latency and less available number of qubits. Thus, a single quantum circuit at a higher level has to be divided into multiple sub-circuits. Due to the high data exchanging overhead, the execution efficiency is severely affected by the number of sub-circuits. Therefore, *dgQuEST* reorders the quantum gates with greedy sub-circuit partitioning based on the constructed DAG to reduce the number of sub-circuits when the memory level is lowered. Specifically, successor nodes has a greater chance of having the same target qubit as its predecessor node. Besides, since the interdependent data introduced by each sub-circuit derived from a DAG must meet the maximum memory capacity, *dgQuEST* also needs to constrain the number of target qubits. Thus, we keep selecting a new gate with no predecessor and applying depth-first search from it to add nodes to the sub-circuit until the threshold of target qubit number is reached or no more nodes can be added.

Figure 3 demonstrates an example of aggregating and partitioning a quantum circuit with 7 qubits in a *three-level memory model*, where the number of qubits within the GPU global memory is 5 and the number of qubits within the shared memory is 2. Different colors indicate the result of partitioning the circuit into sub-circuits at GPU global memory level, and with the same color, different patterns further indicate the result of partitioning at the shared memory level. The actual calculation occurs at the shared memory level (the register level is not considered), before which the data will be migrated from the higher level into the lower level according to the partition result.

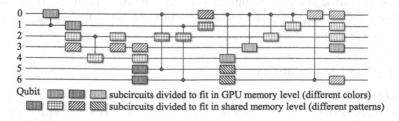

Qubit ▦ ▦ ▦ subcircuits divided to fit in GPU memory level (different colors)
▦ ▦ ▨ ▨ subcircuits divided to fit in shared memory level (different patterns)

Fig. 3. A quantum circuit aggregated and partitioned during 2 memory level lowering.

Intra-node Gate Aggregation and Fusion - For intra-node optimization, *dgQuEST* utilizes shared memory on GPU and applies *gate aggregators* to reorder the execution of quantum gates in two stages to reduce the amount of both global and shared memory access. In the first stage of gate aggregation, the quantum circuit is divided into several sub-circuits by limiting the maximum target number of qubits of the gate aggregator within the shared memory capacity. The second stage of the gate aggregation (a.k.a., *gate fusion*) gathers the quantum gates of the same target qubit by limiting the number of target qubits to 1, which enables computations with continuous targets. These two stages of gate aggregation further improve the data reuse within shared memory and register during execution.

3.4 Remap-Based Data Reorganization

For better locality, we ensure the data dependency of a global-memory-level sub-circuit is within global memory level by remapping qubits to different state vector index bits and reorganizing data. The remap-based data reorganization involves three stages, including **qubit remapping, in-page data reorganization**, and **page remapping**. **Qubit remapping** dynamically remaps the independent qubits into higher memory address bits according to the data dependency obtained by the *gate aggregation*. **In-page data reorganization** rearranges the data within the *offset* bits in the page to adapt the new mapping, which is accelerated by GPU for better performance. **Page remapping** reorganize the data within the *page number* bits in the page by modifying the page table with new mappings for each page. The **qubit remapping** and **in-page data reorganization** are performed after the quantum gate calculation and before sending or writing back the results, while **page remapping** is handled during sending or writing back the results.

Figure 4 demonstrates an example of remap-based data organization with 16 qubits, where the upper 8 bits of state vector index represent page number and the light-colored bits are the qubits without any data dependency. The remapping process can be divided into several steps in 2 scopes: 1) remapping in the scope of the index bits inside a task block within GPU memory, where an in-page data reorganization is needed to keep the consistency between data and the new mapping; 2) remapping in the scope of page number, where page table is changed to match the new mapping. After remapping, qubits involved in the next sub-circuit are all mapped to offset bits inside a GPU task data block and data dependencies are always inside the block.

3.5 Pipelined Communication

In *dgQuEST*, there are the following forms of data migration: 1) Loading a page from the main memory to GPU global memory (*load*); 2) Writing back a page from the GPU global memory to the local node (*write back*); 3) Sending a page from the GPU global memory to a remote node (*send*); 4) A page is received from a remote node and written to the main memory (*receive*).

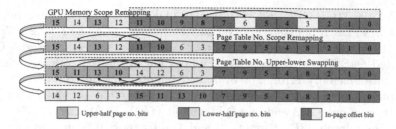

Fig. 4. An example of qubit remapping to state vector index bits. The remapping consists of 3 steps. In-page data is reorganized according to the in-page remapping. Remapping of qubits in page table number scope pages remapping during data exchanging between tasks.

Figure 5 demonstrates the pipelined communication processes. Among them, the *load*, *write back*, and *send* processes of each GPU are controlled by independent threads. The *receive* operation of each node is controlled by a dedicated thread. The *reorganize* operation rearranges the data in memory to match the qubit mapping of the next sub-circuit. Since *write back* occurs more frequently, asynchronous *send* and *write back* operations are adopted to allow overlapping—GPU threads launch asynchronous *send* of a page targeted to other nodes and continue to perform short-running *write back* operations to avoid blocking the execution of the next tasks. Therefore, the followed task execution can overlap with the other three data migration forms of the previous task, which can significantly improve communication efficiency.

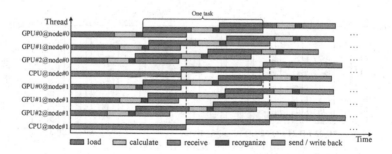

Fig. 5. An execution scenario with data transfer pipelined

4   Evaluation

4.1   Experimental Setup

To evaluate the capability and performance of large-scale quantum circuits simulation by *dgQuEST*, we simulate various circuits with *dgQuEST* on our two-node GPU cluster. Each node consists of two Intel Xeon E5-2680 v4 CPU, two

NVIDIA V100 32 GB GPU and 384 GB DDR4 DRAM. Two nodes are connected
with 40Gb/s FDR Infiniband. For comparison, we choose *QuEST* v3.2.1 [7] and
Qibo v0.1.6.dev2 (with qibotf library v0.0.2) [3]. We did not compare with GPU
accelerated *QuEST* and *HyQuas* as they both fail to run the evaluated circuits
due to out-of-memory error. For evaluated quantum circuit datasets, we use
qubits ranges from 32 to 34 with 1) randomly generated circuits (*RD*), 2) a
combination of one GHZ and one QFT circuit (*GQ*), 3) BV circuit (*BV*), and
4) supremacy circuit (*SP*). Circuits 3) and 4) are exported from Cirq [2].

4.2   Overall Performance

We simulated the above-mentioned quantum circuits with 32 to 34 qubits using
QuEST, *Qibo*, and *dgQuEST* on our GPU cluster. The simulation times and
the speedups against *QuEST* are demonstrated in Fig. 6. Note that *Qibo* can
only run on one node and it fails to simulate circuits with qubits larger than 32
due to out-of-memory error. Both the execution of *QuEST* and *dgQuEST* are
distributed on two nodes. As shown in Fig. 6, *dgQuEST* exhibits 455x (geomean)
speedup on 34 qubits compared to the CPU implementation of *QuEST*, and 120x
(geomean) speedup on 32 qubits compared to *Qibo* on one node. The significant
performance improvement comes from two folds. One is our three-level CPU-
GPU hybrid memory model and corresponding page-based memory management
approaches can effectively utilize the large memory capacity of DRAM with
less swapping overhead, which enables GPU acceleration for larger-scale qubit
simulations. The other is our data reorganization and pipelined communication
can largely reduce both the memory and time overhead for data exchanges.

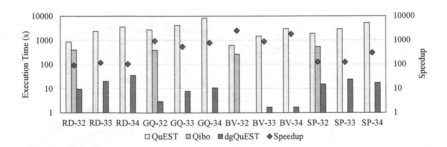

Fig. 6. Overall execution time of *QuEST*, *Qibo*, and *dgQuEST* and the speedup of
dgQuEST against *QuEST*. Both execution times and speedups are shown in log scale.

4.3   Qubit Scalability

To prove the scalability of the number of simulated qubits within a given memory capacity, we simulate GZ quantum circuits with qubits ranging from 31 to
35 on both one node (*1n2g*) and two nodes (*2n4g*) with *dgQuEST*. The execution times of each evaluated circuit with different configurations are shown in

Fig. 7, where the execution times increase linearly when the simulated number of qubits increases with both one node and two nodes execution. For more detailed comparisons, we analyze the memory usage of the state vector, *QuEST*, and *dgQuEST* during distributed simulations of 31–35 qubits, as shown in Fig. 8. The results demonstrate that *dgQuEST* sharply reduces the memory requirements of simulating distributed on multiple nodes by nearly two folds due to the page-table based memory management. The memory used by *dgQuEST* is nearly equal to the required memory for the state vector. Therefore, the saved large amount of memory further contributes to simulating more qubits within the same memory capability.

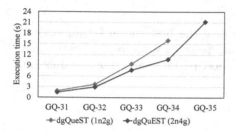

Fig. 7. Execution time and speedup of *dgQuEST* simulating GQ with different number of qubits.

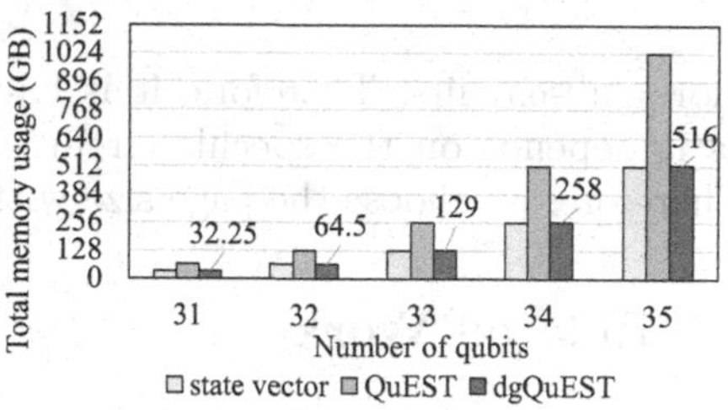

Fig. 8. Evaluated memory usage of state vector, *QuEST*, and *dgQuEST* simulating different numbers of qubits.

4.4   Strong Scalability

To demonstrate the strong scalability of *dgQuEST*, we compared the execution time of each circuit simulation on one node and two nodes. As shown in Fig. 9, most of the evaluated simulations exhibits poor or even negative scalability when it distributes to more nodes. Although the *gate aggregation* can minimize the data exchanges between nodes and *pipelined communication* enables overlapping of computations and communications, the data exchange still becomes a bottleneck when the quantum circuit is not deep enough (e.g., only 79 gates in *BV-32*). Therefore, our approaches to reduce the distributing memory overhead as well as support more qubits regardless of the limited GPU memory capacity can significantly reduce the required nodes for each quantum simulation and thus contribute to overall performance improvement.

4.5   Sensitivity Analysis on Page Size

The page size affects the performance of *dgQuEST*, as shown in Fig. 10. In general, a larger page size leads to a larger task block and potentially fewer data exchanges, while it makes the execution of each task longer and results in less pipeline utilization. In contrast, with a smaller page size, each execution takes less time and the data exchange pipeline is more utilized, but it will result in

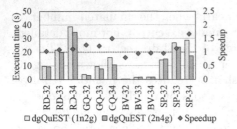
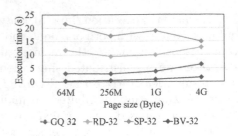

Fig. 9. Execution time and speedup of *dgQuEST* with one node (*1n2g*) and two nodes (*2n4g*).

Fig. 10. Execution time of *dgQuEST* simulating 32-qubit circuits on two nodes with different page sizes.

more sub-circuits. Therefore, it is hard to give the best choice for all circuits as it depends on the specific circuit to tune page sizes for best performance. Therefore, we choose the page size with the best performance in our evaluation.

5   Related Work

Several optimizations have been proposed for the full-amplitude quantum simulation. qHipster [10] optimizes the performance with SIMD, thread allocation, communication, multi-threading, and cache blocking. JUQCS [11] provides the GPU accelerated implementation, which reduces communication overhead by relabeling global and local qubits. *Qibo* [3] utilizes TensorFlow and custom operators for its backend and supports acceleration with multiple GPUs. *HyQuas* [12] utilizes both shared memory and batch matrix-vector multiplication to accelerate the quantum simulation, and it implements muiti-GPU simulation and optimizes the communication between GPU. However, *HyQuas* fails to exploit the hybrid CPU-GPU memory hierarchies to support large-scale quantum circuit with more qubits on limited number of nodes. Whereas, *Qibo* cannot utilize distributed GPUs across multiple nodes. Different from above approaches, *dgQuEST* leverages the memory capacity of both CPUs and GPUs across distributed nodes for large-scale quantum circuit simulations with accelerated performance.

6   Conclusion

In this paper, we propose *dgQuEST*, a novel acceleration method that utilizes the CPU-GPU hybrid memory hierarchies for large-scale quantum circuit simulation. *dgQuEST* adopts efficient memory management and communication schemes for better memory access performance and reduced communication overhead across distributed GPU nodes. The experiment results demonstrate that compared to *QuEST*, *dgQuEST* can achieve an average speedup of 403× when simulating quantum circuit with 32 qubits. In addition, *dgQuEST* can easily scale to quantum circuit simulation with 35 qubits on two distributed GPU nodes, far beyond the state-of-the-art GPU implementation such as *HyQuas* can support.

Acknowledgements. This work was supported by National Key Research and Development Program of China (No. 2020YFB1506703), National Natural Science Foundation of China (No. 62072018), and State Key Laboratory of Software Development Environment (No. SKLSDE-2021ZX-06). Hailong Yang is the corresponding author.

References

1. Cai, Z.: Multi-exponential error extrapolation and combining error mitigation techniques for NISQ applications. NPJ Quant. Inf. **7**(1), 80 (2021). https://doi.org/10.1038/s41534-021-00404-3
2. Developers, C.: Cirq (2021),.https://doi.org/10.5281/zenodo.4750446. See full list of authors on Github: https://github.com/quantumlib/Cirq/graphs/contributors
3. Efthymiou, S., et al.: Qibo: a framework for quantum simulation with hardware acceleration (2020)
4. Endo, S., Benjamin, S.C., Li, Y.: Practical quantum error mitigation for near-future applications. Phys. Rev. X **8**, 031027 (2018). https://doi.org/10.1103/PhysRevX.8.031027
5. Gray, J.: quimb: A python package for quantum information and many-body calculations. J. Open Source Softw. **3**(29), 819 (2018). https://doi.org/10.21105/joss.00819
6. Huang, C., Szegedy, M., Zhang, F., Gao, X., Chen, J., Shi, Y.: Alibaba cloud quantum development platform: applications to quantum algorithm design (2019)
7. Jones, T., Brown, A., Bush, I., Benjamin, S.C.: Quest and high performance simulation of quantum computers. Sci. Rep. **9**(1), 10736 (2019). https://doi.org/10.1038/s41598-019-47174-9
8. McArdle, S., Jones, T., Endo, S., Li, Y., Benjamin, S.C., Yuan, X.: Variational ansatz-based quantum simulation of imaginary time evolution. NPJ Quant. Inf. **5**(1), 75 (2019). https://doi.org/10.1038/s41534-019-0187-2
9. McCaskey, A.J.: Quantum virtual machine (qvm). version 00 (2016). https://www.osti.gov/biblio/1339996
10. Smelyanskiy, M., Sawaya, N.P.D., Aspuru-Guzik, A.: qhipster: The quantum high performance software testing environment (2016)
11. Willsch, D., Willsch, M., Jin, F., Michielsen, K., Raedt, H.D.: Gpu-accelerated simulations of quantum annealing and the quantum approximate optimization algorithm (2021)
12. Zhang, C., Song, Z., Wang, H., Rong, K., Zhai, J.: Hyquas: hybrid partitioner based quantum circuit simulation system on GPU. In: Proceedings of the ACM International Conference on Supercomputing, ICS '21, pp. 443–454. Association for Computing Machinery, New York (2021). https://doi.org/10.1145/3447818.3460357

vSketchDLC: A Sketch on Distributed Deep Learning Communication via Fine-grained Tracing Visualization

Yanghai Wang, Shuo Ouyang, Dezun Dong$^{(\boxtimes)}$, Enda Yu, and Xiangke Liao

College of Computer, National University of Defense Technology, Changsha, China
{wangyanghai,ouyangshuo,dong,yuenda,xkliao}@nudt.edu.cn

Abstract. Intensive communication cost for gradients and parameters is becoming the bottleneck of distributed deep learning training. It is crucial for optimizing such communication bottleneck through measuring communication operations effectively. However, many existing communication measurement tools, such as MXNet profiler, still suffer from serious limitations. Specifically, they cannot satisfy two requirements simultaneously, that is, fine-grained collection of low-level communication operations and user-friendly analysis of comprehensive measurement results. In this paper, we make the first attempt to propose an open-sourced, fine-grained and user-friendly communication measurement tool on top of MXNet, called vSketchDLC. vSketchDLC can trace low-level communication events between framework and communication library interface, and capture end-to-end push and pull communications between workers and servers. It supports to generate communication records in standard format, enabling users to analyze the communication traces by merely using standard visualization tools such as Chrome Trace Viewer. Our design exploits in-memory buffers and asynchronous record writes to ensure measurement activities do not impact training performance. We conduct extensive experiments on a public-cloud GPU cluster to verify the effectiveness of vSketchDLC for MXNet. Experimental results show that vSketchDLC can empower users to analyze fine-grained communication records through friendly interactions, and identify potential training bottlenecks from multiple perspectives, including training timeline and iterations, DNN layers, workers or servers, etc. We can observe the relationship between different communications visually, i.e., to highlight a selected period of communication traces, to zoom in or zoom out, such that identifying the root causes of communication bottleneck and seeking to improve training performance.

Keywords: Communication measurement tool · Distributed deep learning · Parameter server · MXNet

---

Y. Wang and S. Ouyang—Equal contribution.

1   Introduction

Distributed deep neural network (DNN) training suffers from high communication cost caused by frequent data synchronization, especially in the paradigm of data parallel. Recent years, there have been tremendous studies focus on communication bottleneck optimization from various perspectives [9], such as communication scheduling [12] and communication frequency reduction [17]. These efforts alleviate the communication bottleneck in distributed training and improve the training speed to varying degrees.

However, we observe that tons of existing studies only provide a coarse-grained analysis of the communication optimization [12,17]. Most studies take end-to-end training speed (as known as training throughput, single iteration or epoch time) as the only metric for performance comparison, which hides a lot of low level communication details. It is difficult to reflect which communication leads to bottlenecks or whether there still remains potential bottlenecks after optimization. Therefore, a fine-grained communication measurement tool for general deep learning platform is necessary.

A fine-grained measurement tool should (i) be capable for recording network communication cost correctly and precisely, (ii) occupy computation and communication resources as few as possible and (iii) be easy to analyze its measurement results. First, the main purpose of using measurement tool is locating the communication bottleneck of distributed DNN training. Therefore, a detailed communication time should be provided for the front-end user under the premise of ensuring correctness and precise. If the tool does not expose much detailed information, the user cannot locate and optimize the bottleneck. Second, since the distributed DNN training requires many computation (CPU, host memory and GPU memory) and communication (network bandwidth) resources, the training system can only provide limited resources for measurement tool. What the measurement tool needs to do is to capture the communication information without affecting the training performance, which requires it to be efficient. Third, a qualified measurement tool should be easy to analyze its measurement results. In other words, researchers can simply use the measurement results to analyze the communications comprehensively.

Unfortunately, many existing communication measurement tools can not achieve all of these three visions. Popular deep learning frameworks, such as TensorFlow [2], PyTorch [11] and MXNet [3], provide profiler tools to capture the execution time of operators in computation graphs and save these information to files in standard format, which can be visualized and analyzed through Chrome Trace Viewer [4]. However, these tools are developed for computation graphs and only record the operator level information, and communication related operations are usually treated as an operator in computation graph. Hence, records captured by these tools lack communication details, and are difficult to carry out fine-grained communication analysis. Unlike above frameworks, Horovod [14], which uses collective communication verbs (such as allreduce and allgather) for intra- and inter-node communication, implements a tool named Timeline. Timeline can record fine-grained communication, but it only serves point-to-point communication architecture. Other communication measurement tools, such as SketchDLC [16], can record communication details for parameter server architecture in MXNet, but its records cannot be visualized, so it cannot display the overall communication sketch in training. In summary, it is worth to develop a better communication measurement tool for deep learning frameworks using parameter server architecture.

In this paper, we make the first attempt to propose an open-sourced, fine-grained and user-friendly communication measurement tool vSketchDLC [1] in deep learning framework MXNet. This tool can capture fine-grained communication records, and can benefit from existing visualization tools such as Chrome Trace Viewer to facilitate analyzing the communication behaviours. We summarize our main contributions as follows: (i) We catch the low-level fine-grained communication operations and generate record files in standard format to display them through Chrome Trace Viewer. (ii) After visualizing communication records, we provide potential training bottlenecks from multiple perspectives and give possible solutions. (iii) We improve the method of generating record files, i.e., making vSketchDLC do not affect the training performance when writing record files.

The rest paper is organized as follows. We introduce the communications in MXNet and discuss our motivation in Sect. 2 and Sect. 3. Then we present our design and implementation of vSketchDLC in Sect. 4, followed by the experiments and conclusion in Sect. 5 and Sect. 6.

2   Background

2.1   Communication System of MXNet

The communication system of MXNet is divided into three levels [8]. (i)*KVStore* first merges the computing results on different local devices and stores them in key-value pairs, equivalent to a distributed database, which store the data that will be communicated. (ii)Then ps-lite slices the merged results of *KVStore* into data segments, and will send them to different servers with communication interface. (iii)Finally, ZeroMQ [6] interface transmits the data segments in ps-lite to complete the communications. It is notable that we can capture communications at diverse levels, but the granularity of communications we obtain is different. Capturing communications from lower levels can catch more communication details.

In *KVStore*, we can simply capture operator-level communication records. The communication operations in *KVStore* are based on the communication operators in computation graphs of MXNet, and are mainly divided into push operator and pull operator. We can record the start time and the end time of executing these operators. In ps-lite, we can capture fine-grained low-level communication operations inside communication operators. When executing a communication operator in ps-lite, we actually perform two network transmissions and one server message processing. Dividing the communication operator into fine-grained communication operations can help us better find potential training bottlenecks. In ZeroMQ, we can also capture fine-grained communications. Nevertheless, it is difficult to capture the relationship between the low-level communication operations of a communication operator in ZeroMQ. In summary, it seems a better choice to capture communications in ps-lite.

2.2   Communication Measurement Tools in MXNet

MXNet provides its communications measurement tool profiler. It captures communications in *KVStore*, and generates communications records in standard format. Some other works are devoted to capturing communication sketch at low level, such as SketchDLC, which captures communication in ps-lite. These tools have their own advantages.

MXNet profiler can generate communication records in standard format. This allows profiler to display visual communication records with the existing visualization tools such as Chrome Trace Viewer, which can view the communications in a certain time period, find out all communication operators belonging to a training iteration, or select some communication operators for analysis. Meanwhile, Trace Viewer is able to zoom in and out the visual communication records, which enables us to understand the communication process in MXNet at a certain degree.

SketchDLC captures fine-grained communication records in MXNet, and can analyze specific communications in depth. It can find out specific communication records for analysis, for instance, the communications belonging to a DNN layer, a training iteration, a worker or server, etc. SketchDLC can also analyze communications from other aspects, such as the overlap rate of computation and communication in training process, which helps us to locate training bottlenecks.

In experiments, these existing communication measurement tools can greatly improve our understanding of the communications for distributed training in MXNet, making it easier to find out training bottlenecks.

3   Motivation

We hope to analyze training bottlenecks resulted from communications of MXNet in such steps. First, capturing fine-grained low-level communication operations. Second, making full use of existing visualization tools to analyze communication records from multiple perspectives, such that greatly simplifying the analysis process and the difficulty of finding training bottlenecks. However, existing communication measurement tools in MXNet cannot satisfy our requirements.

MXNet profiler provides only operator-level communication records, and it is difficult to explore many valuable informations from its visual communication records generated by Chrome Trace Viewer. The profiler simply uses the start time and end time of communication operators to display the execution of all communication operators, and give no basis for dividing communications, such

(a) zoom in   (b) zoom out

Fig. 1. We use Chrome Trace Viewer to zoom in and out the communication records captured by vSketchDLC. All communication records are divided by the DNN layer to which they belong, and the server that processes this communication. Figure 1a shows the communications in three training iterations and six DNN layers. Figure 1b zooms out the communication records and shows the communications in more training iterations and ten DNN layers, which can help us find the relationship between different communications.

as DNN layers. This leads to overlapping visual communication operators, and can only illustrate whether there are communications within a time period from the time perspective.

SketchDLC is capable to analyze captured fine-grained communication records from multiple perspectives. Unfortunately, it has to pay high time costs to discover training bottlenecks. SketchDLC cannot use existing visualization tools, i.e., it has to seek out needed communication records with specific features from the mass of record files, which requires unbearable search time. Moreover, it cannot display the relationship between different communications visually, which means that it has to analyze more relevant communications to find training bottlenecks. These factors make the process of analyzing training bottlenecks unbearable time costs.

In short, we want to propose such a communication measurement tool in MXNet to satisfy our requirements. It can not only capture fine-grained communication records, but also provide user-friendly interactions provided by Chrome Trace Viewer to quickly and comprehensively analyze communication records. Due to the intuitiveness of visual records, we can understand the communication process of MXNet, observe the relationships between different communications, and identify potential training bottlenecks. In addition, we put forward a new requirement for this tool, it should solve performance problems existing in MXNet profiler and SketchDLC, in other words, not affect training performance when generating record files. In the end, we propose vSketchDLC to achieve our requirements.

4   Design and Implementations

4.1   Analyzing Communications with vSketchDLC

vSketchDLC allows us to identify training bottlenecks with following two user-friendly ways. We show an example of fine-grained communication records generated by vSketchDLC in Fig. 1, which is visualized by Chrome Trace Viewer. In

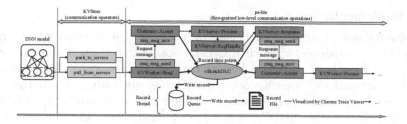

Fig. 2. In ps-lite, when executing a communication operator of MXNet, the worker first sends a request to the server and then receives the response from it. vSketchDLC captures low-level communication operations in this process, and starts a asynchronous record thread on each worker to write files and avoid affecting training performance. In ps-lite, when executing a communication operator of MXNet, the worker first sends a request to the server and then receives the response from it. vSketchDLC captures low-level communication operations in this process, and starts a asynchronous record thread on each worker to write files and avoid affecting training performance.

example, we use synchronous distributed training mode to train a DNN model. In this mode, a training iteration must wait for the end of all communications in the previous training iteration to start.

Observation. Since vSketchDLC exposes the relationships between different communications, we can directly observe potential training bottlenecks. In example, the communications in DNN layer marked as Process–26 have longer execution time and latter end time, and are more likely to affect training performance. Moreover, vSketchDLC illustrates that the communications on different servers may be unbalanced. The communications of the DNN layer marked as Process –22, –24, –26, are mainly processed by server 1, this imbalance may delay the end of communications of these DNN layer. We can also explore potential training bottlenecks from other perspectives, such as evaluating the proportion of network transmission time in executing communication operators.

Selection. Since vSketchDLC generates communication records in a reasonable way, Chrome Trace Viewer can provide user-friendly interactions to help us analyze communications from multiple perspectives. We can use the mouse to select a period of communication records, which may belong to different DNN layers, different workers, different servers, and different training iterations, etc., then Trace Viewer will provide total execution time, average execution time and other informations of all the same name records in this period. We give different names to the low-level communication operations in push and pull operators, so we can analyze them respectively. In addition, Trace Viewer allows us to zoom in and out visual communication records, hence we can select communication records in different time periods until the entire training process, or some DNN layers until all DNN layers.

4.2   Implementations

We solve following four main challenges in ps-lite to achieve vSketchDLC. After solving these challenges, vSketchDLC can capture fine-grained standard communication records in ps-lite to benefit from existing visualization tools, and record comunications on each worker asynchronously with a record thread to avoid affecting training performance. We show the process of vSketchDLC in Fig. 2. To achieve vSketchDLC, we first have to understand the running process of ps-lite. ps-lite performs four key functions in sequence to complete a communication operator in MXNet: the worker calls *Send* function to send a request message, and the server calls *Accept* function to receive this request. After message processing, the server calls *Response* function to send a response, and the worker calls *Accept* function to receive this response.

Clock Inconsistency. We record time points at the above four key functions in ps-lite, but the clocks on different computing nodes are inconsistent, and it is difficult to record accurate time points. Targeting this issue, we take the time points recorded on the same computing node to compute time intervals. We adopt the time points recorded in *Send* function and *Accept* function on the worker to compute the execution time of communication operators, the time points recorded in *Accept* function and *Response* function on the server to compute server processing time, and finally use the execution time minus processing time to compute network transmission time. In this way, we can avoid generating incorrect communication records resulted from clock inconsistency.

Message Dependency. In ps-lite, after the worker sends a request to the server, the server will send back the corresponding response. Unfortunately, since a worker sends multiple requests and receive multiple responses in each training iteration, it is difficult to determine the dependencies between requests and responses. To solve this problem, we find a feature of MXNet communication process. When communicating a DNN layer in a training iteration, a worker only sends one request to each specific server and receives a response from each server. Therefore, we can identify the dependency between a request and a response by determining the DNN layer and the server that processes the request. We enable vSketchDLC to capture the DNN layer information from MXNet, and obtain server rank from message meta in the response message.

Incomplete Record. We have to record complete communication operators into one file for correctly visualizing communication records. Since we record time points on workers and servers, we can write records files on both of them, but it is difficult to lock threads on different computing nodes, which will lead to incomplete communication records. Therefore, we choose to write record files only on workers. Before the server sends the response, we save the required information from the request received by the server to the response, such that we can obtain complete informations to generate communication records from the response on the worker.

Appropriate Visualization. In order to observe the relationship between different communications more easily and use the existing visualization tools more effectively, we should generate communication records in an appropriate way. Therefore, we first divide all communication records by DNN layers, then divide the records belonging to a DNN layer by servers, finally display communication records in chronological order. In this way, we can use visualization tools to provide user-friendly analysis of communications from multiple perspectives.

5 Experiments

Due to the limitations of the experimental environment, we use two or four computing nodes, but actually vSketchDLC can be applied to larger clusters. We adopt synchronous distributed training mode to conduct our experiments. In this section, we analyze training bottlenecks from three aspects and give possible optimization methods. We first use vSketchDLC to illustrate the communications on different workers respectively. Then we compute the time composition of communication operators when training different DNN models. Finally, we analyze the communications on different servers when using multiple servers for training.

5.1 Experimental Configurations

Datasets and DNN Models. We employ ImageNet 1K [13] dataset that has about 1.3 million training images and 50000 validation images, and conduct experiments with AlexNet [7], VGG16 [15], ResNet50 [5] models.

Infrastructures. We use distributed clusters in PARATERA [10], PARATERA is a public cloud service provider focused on high performance computing. Each computing node in clusters has two E5-2660 10C V3 Xeon CPUs with 256 GB of RAM, equips two NVIDIA Tesla K80 cards and a 56 Gbps IB network card, and are installed with 64-bit CentOS 7.2.1511 with CUDA Toolkit 8.0 and cuDNN 6.0. We conduct experiments with MXNet 1.4.1, Open MPI 2.1.6, and implement vSketchDLC in ps-lite. We use IPoIB protocol to conduct our experiments.

5.2 Communications on Different Workers

Since MXNet allows us to start a server and a worker on a computing node simultaneously, we use vSketchDLC to explore the difference in communications between different workers in following two cases. (i) Using two computing nodes, one node starts a server and a worker, another node starts a worker. (ii) Using three computing nodes, one node starts a server, other two nodes start a worker. We set batchsize as 256, train the AlexNet model, and show the experiment results in Fig. 3. Each communication operator is divided into low-level network transmission time and server processing time, and displayed according to the DNN layer.

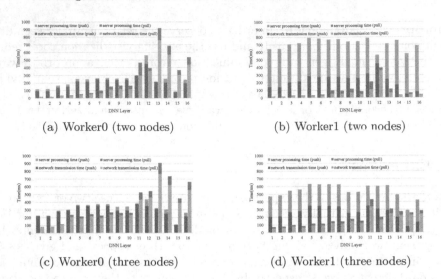

(a) Worker0 (two nodes)   (b) Worker1 (two nodes)

(c) Worker0 (three nodes)   (d) Worker1 (three nodes)

Fig. 3. In each training iteration, a DNN layer performs push and pull two kinds of communication operators. vSketchDLC records the average execution time of the low-level communication operations of these communication operators in each iteration.

Under different experimental configurations the communications on workers are quite different. First, when using three nodes, the time difference of communicating DNN layers between different workers is smaller than using two nodes. It may be related to the fact that the server does not need to use the network card when communicating with the worker on the same node, hence starting a server and a worker on the same computing node may lead to larger communication differences between different workers. Second, the server processing time varies greatly among push operators on different workers, which may be caused by a worker becoming a training bottleneck. In synchronous mode, at a training iteration, the server have to wait for the gradients of this DNN layer from all workers to be collected before sending a response, then workers can pull parameters from servers. Since vSketchDLC records the server waiting time into the server processing time, if some workers push the gradients late, there will be a huge server processing time. In summary, we may alleviate training bottlenecks from two aspects: starting the server individually on a computing node to mitigate the imbalance between communications of different worker, or check whether there are slower computing nodes in the cluster.

5.3   Time Composition of Communication Operators

We train AlexNet, VGG16, ResNet50 models and use vSketchDLC to evaluate their requirements for network performance and server processing time. To facilitate data statistics, we use two or four computing nodes, start a server and a worker on one node, and start a worker on other nodes. The batchsize is set as

256 when using two nodes and 512 when using four nodes. Since the communications on different workers are different, we use the communication records on all workers to compute the average execution time of low-level communication operations. We show the experiment results in Fig. 4.

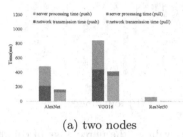
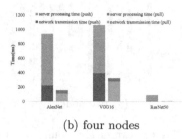

(a) two nodes   (b) four nodes

Fig. 4. We train different DNN models, and compute the average time of network transmission and server processing in all push and pull operators to analyze the time composition of communication operators.

We get some observations through analyzing the time composition of executing communication operators. (i) First, VGG16 has the highest requirement for network performance, followed by AlexNet. When training ResNet50, communications almost constitute no training bottlenecks. This can guide us to effectively avoid training bottlenecks. For instance, using ResNet50 can suffer less training performance loss on clusters with lower network performance. (ii) Second, when the number of nodes increases, the server processing time in push operator increases rapidly, while the network transmission time and the execution time of pull operator changes little. The possible reason is that some workers have become training bottlenecks in synchronous mode, we explain it in the previous section. When the cluster size increases, the average server processing time increases more obviously due to more delayed workers, so we can exclude these slower computing nodes to improve training performance.

5.4   Communications on Multiple Servers

In experiments, we observe following three phenomena. (i) If a DNN layer is split and sent to different servers, the push operators almost start simultaneously, then the worker must wait for all push operators to finish before starting the pull operators. (ii) A training iteration must wait until all pull operators are completed. (iii) The same DNN layer on different workers is always split and communicated to the same server. These factors mean that if one communication between a worker and a server is longer than other servers, it may produce training bottlenecks. Therefore, we evaluate the communications on different servers according to DNN layers, We train the AlexNet model, use two or four computing nodes, starts a server and a worker on each node, set batchsize as 256 when using two nodes and 512 when using four nodes, take the sum of the execution time of the push and pull operators as the communication time between the worker and a server. We use the communication records on all workers to compute the experimental results, and show them in Fig. 5.

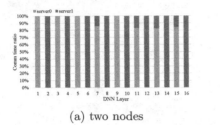

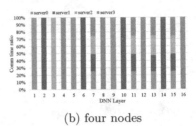

(a) two nodes   (b) four nodes

Fig. 5. We use two or four computing nodes, and evaluate the percentage of communication time on different servers when communicating a DNN layer. When the DNN layer does not communicate with a server, the percentage of communication time on this server is 0.

In experiments, the communications on different servers varies greatly. If the DNN layer is split, when using two nodes, server0 will suffer huge communications, while server1 will be idle. Instead, the proportion of communication on different servers is close when using four nodes. This means that MXNet may encounter training bottlenecks due to unbalanced communications on different servers when using a small amount of computing nodes. To solve this problem, we may find a better communication balance strategy to make MXNet balance the communication time on different servers in more cases.

6   Conclusion

MXNet is one of the most popular deep learning frameworks, but existing communication measurement tools in MXNet cannot generate fine-grained communication records and benefit from visualization tools simultaneously. We first proposed vSketchDLC to ahieve these two requirements. vSketchDLC records low-level communication operations of communication operators in MXNet, then use visualization tool to help us analyze communications comprehensively and discover training bottlenecks from multiple perspectives. In experiments, we analyze the communications on workers or servers, and the time composition of executing the communication operators in MXNet to discover potential training bottlenecks.

Acknowledgment. This work is supported by the National Key R&D Program of China (Grant No.2018YFB0204300), Excellent Youth Foundation of Hunan Province (Dezun Dong) and National Postdoctoral Program for Innovative Talents under Grant No. BX20190091.

References

1. vSketchDLC. GitHub repository. https://github.com/HiNAopen/vSketchDLC
2. Abadi, M., et al.: Tensorflow: a system for large-scale machine learning. In: 12th USENIX Symposium on Operating Systems Design and Implementation (OSDI), pp. 265–283 (2016)
3. Chen, T., et al.: Mxnet: a flexible and efficient machine learning library for heterogeneous distributed systems. arXiv preprint arXiv:1512.01274 (CoRR) (2015)
4. Google: Trace viewer. GitHub repository. https://github.com/catapult-project/catapult/tree/master/tracing
5. He, K., Zhang, X., Ren, S., Sun, J.: Deep residual learning for image recognition. In: IEEE Conference on Computer Vision and Pattern Recognition (CVPR), pp. 770–778 (2016)
6. Hintjens, P.: Zeromq: the guide. http://zguide.zeromq.org/page:all
7. Krizhevsky, A., Sutskever, I., Hinton, G.E.: Imagenet classification with deep convolutional neural networks. In: Advances in Neural Information Processing Systems (NIPS), pp. 1097–1105 (2012)
8. Li, M., et al.: Improving the performance of distributed mxnet with rdma. Int. J. Parallel Program. **47**, 467–480 (2019)
9. Ouyang, S., Dong, D., Xu, Y., Xiao, L.: Communication optimization strategies for distributed deep neural network training: a survey. J. Parallel Distrib. Comput. **149**, 52–65 (2021)
10. PARATERA: Paracloud. https://cloud.paratera.com
11. Paszke, A., et al.: Pytorch: an imperative style, high-performance deep learning library. In: Advances in Neural Information Processing Systems (NIPS), pp. 8026–8037 (2019)
12. Peng, Y., et al.: A generic communication scheduler for distributed DNN training acceleration. In: Proceedings of the 27th ACM Symposium on Operating Systems Principles (SOSP), pp. 16–29 (2019)
13. Russakovsky, O., et al.: Imagenet large scale visual recognition challenge. Int. J. Comput. Vis. **115**, 211–252 (2015)
14. Sergeev, A., Del Balso, M.: Horovod: fast and easy distributed deep learning in tensorflow. arXiv preprint arXiv:1802.05799 (CoRR) (2018)
15. Simonyan, K., Zisserman, A.: Very deep convolutional networks for large-scale image recognition. In: 3rd International Conference on Learning Representations (ICLR) (2015)
16. Xu, Y., Dong, D., Xu, W., Liao, X.: SketchDLC: a sketch on distributed deep learning communication via trace capturing. ACM Trans. Arch. Code Optim. (TACO) **16**, 1–26 (2019)
17. Xu, Y., Dong, D., Zhao, Y., Xu, W., Liao, X.: OD-SGD: one-step delay stochastic gradient descent for distributed training. ACM Trans. Arch. Code Optim. (TACO) **17**, 1–26 (2020)

Scalable Algorithms Using Sparse Storage for Parallel Spectral Clustering on GPU

Guanlin He[1,2]([⊠]), Stephane Vialle[1,2], Nicolas Sylvestre[1], and Marc Baboulin[1,3]

[1] Université Paris-Saclay, CNRS, LISN, Orsay 91405, France
`guanlin.he@lisn.fr`, {`nicolas.sylvestre,marc.baboulin`}`@upsaclay.fr`
[2] CentraleSupélec, Gif-sur-Yvette 91192, France
`stephane.vialle@centralesupelec.fr`
[3] Université Paris-Saclay, CNRS, CEA, Maison de la Simulation,
Gif-sur-Yvette 91191, France

Abstract. Spectral clustering has many fundamental advantages over k-means, but has high computational complexity ($\mathcal{O}(n^3)$) and memory requirement ($\mathcal{O}(n^2)$), making it prohibitively expensive for large datasets. In this paper we present our solution on GPU to address the scalability challenge of spectral clustering. First, we propose optimized algorithms for constructing similarity matrix directly in CSR sparse format on the GPU. Next, we leverage the spectral graph partitioning API of the GPU-accelerated nvGRAPH library for remaining computations especially for eigenvector extraction. Finally, experiments on synthetic and real-world large datasets demonstrate the high performance and scalability of our GPU implementation for spectral clustering.

Keywords: Spectral clustering · GPU computing · Similarity matrix construction · Sparse matrix format · Parallel code scalability

1 Introduction and Positioning

Data Clustering. Also known as cluster analysis, data clustering refers to an automatic process that discovers the natural groupings (i.e. clusters) of a set of unlabeled data instances [7]. It belongs to unsupervised machine learning and is one of the most important and challenging tasks in data analysis and pattern recognition. Generally, the clustering process seeks to maximize intra-cluster similarity and minimize inter-cluster similarity.

Various kinds of approaches to clustering have been proposed in the literature. The most well-known one might be the k-means algorithm, which tries to minimize intra-cluster distance iteratively. Although k-means has the virtue of simplicity and speediness, it only forms convex clusters, as shown in Fig. 1a & 1c. Besides, k-means usually suffers from the "curse of dimensionality" because the Euclidean distance metric typically used in the k-means algorithm will lose sensitivity in high-dimensional space [2,6]. Another disadvantage of k-means is

C. Cérin et al. (Eds.): NPC 2021, LNCS 13152, pp. 40–52, 2022.
https://doi.org/10.1007/978-3-030-93571-9_4

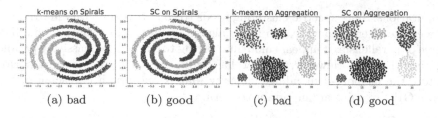

k-means on Spirals	SC on Spirals	k-means on Aggregation	SC on Aggregation
(a) bad	(b) good	(c) bad	(d) good

Fig. 1. k-means vs. spectral clustering (SC) on 2D shape datasets

the sensitivity to randomized centroid initialization with respect to the result of clustering. Consequently, k-means often gets stuck in local minima solutions, and sometimes even generates arbitrarily bad clusterings, as shown in Fig. 1c. A better centroid initialization approach is the k-means++ seeding method, which chooses centroids by adaptive probabilistic sampling and generally improves both the accuracy and the speed of k-means [1].

Spectral Clustering. A more recent clustering method with many fundamental advantages over k-means is spectral clustering [10]. Based on graph theory, it has a close connection with spectral graph partitioning which tries to minimize the volume of connections between clusters relatively to their size, also known as minimizing balanced cut [12]. Essentially, spectral clustering embeds data into the sub-eigenspace of graph Laplacian and then performs k-means on the embedded representation.[1] However, contrary to k-means, spectral clustering can discover non-convex clusters and is more likely to find the global minimum owing to the *embedding* step, as shown in Fig. 1b & 1d. Moreover, as the *embedding* step projects data from $\mathbb{R}^d$ to $\mathbb{R}^{k_c}$, it can play a role of dimensionality reduction for high-dimensional data that has d dimensions and k_c clusters with $d > k_c$, which will benefit the following k-means step. Additionally, when k_c is unknown, the eigenvalues and eigenvectors calculated in spectral clustering algorithm can be exploited to estimate the natural k_c [10,16,18].

Spectral clustering is attractive with the above features, but the algorithm has an important disadvantage: $\mathcal{O}(n^3)$ time complexity [17], mainly due to the calculation of eigenvectors ($\mathcal{O}(n^3)$ when using direct methods) and the construction of similarity matrix ($\mathcal{O}(n^2d)$), for a dataset with n instances in d dimensions. Thus, spectral clustering will have a prohibitive computational cost as n grows. On the other hand, storing the similarity matrix and the graph Laplacian matrix both need $\mathcal{O}(n^2)$ memory space. Therefore, the high complexities of both computational and memory space requirements lead to a great challenge when addressing large datasets with spectral clustering.

One way to meet the scalability challenge of spectral clustering is to employ modern parallel architectures, such as multi-core CPU and many-core GPU. The CPU can run a few dozen heavy threads in parallel, while the GPU can run thousands of light threads in parallel and achieve a higher overall instruction rate and memory bandwidth. Thus, the GPU is specialized for highly parallel

---

[1] Therefore, spectral clustering may also be regarded as the combination of a heavy *preprocessing* step (including main computations) and a classical k-means step.

computations. Since spectral clustering needs to construct similarity matrix, and to calculate eigenvectors through linear algebra computations, both with a high degree of parallelism, it appears more interesting to exploit the massively parallel nature of the GPU. However, the GPU has limited global memory resources. How to store the memory-demanding similarity matrix and graph Laplacian matrix on the GPU remains an important concern.

Related Works. There are existing studies related to GPU-accelerated spectral clustering, but we observed the following limitations. First, the benchmark datasets are usually of limited size [8,19]. Second, many studies [3,4,11] are oriented to spectral graph partitioning instead of spectral clustering, thus they typically assume the edge list or the adjacency list of a graph is available, ignoring the construction of similarity matrix that would take $\mathcal{O}(n^2 d)$ arithmetical operations in the general case of spectral clustering. Third, two research works have reported some limitations in the speedup of similarity matrix construction [19] and eigenvector computation [8]. Therefore, although adapted to the GPU architecture, a particular attention should be paid to the parallelization of these two calculation steps. An example of video segmentation through spectral clustering in pixel level has been successfully implemented on a cluster of GPUs [14] but unfortunately the authors introduced too briefly their parallelization details and did not give performance analysis of their parallel implementation.

Positioning. In this paper, we focus on the parallelization of spectral clustering algorithm on the GPU in order to address large datasets. The main contributions are our optimized parallel algorithms on the GPU for constructing similarity matrix directly in Compressed Sparse Row (CSR) format. This can achieve significant performance improvements, reduce substantial memory space requirement on the GPU, and make it possible to take advantage of the GPU-accelerated nvGRAPH library for subsequent computations of spectral clustering. Moreover, we analyze the effectiveness and performance of eigensolver-embedded algorithms in the nvGRAPH library.

The remainder of this paper is organized as follows. Section 2 reviews the principles of spectral clustering. Section 3 illustrates our solutions and optimized parallel algorithms for similarity matrix construction directly in CSR format on the GPU. Then we present in Sect. 4 our analysis and exploitation of the nvGRAPH eigensolver-embedded algorithms. Finally we conclude in Sect. 5.

2   Spectral Clustering Principles

Given a set of data instances $X = \{x_1, ..., x_n\}$ with x_i in $\mathbb{R}^d$ and the number of desired clusters k_c, the first step of spectral clustering is to construct similarity graph and generate corresponding similarity matrix (see Algorithm 1). Two things are worth noting as they can essentially affect the final clustering result. **(1) How to measure the distance or similarity between two instances.** There are a number of metrics, such as Euclidean distance, Gaussian similarity, cosine similarity, etc. The choice of metric should depend on the domain the data comes from and no general advice can be given [10]. The most commonly used metric seems to be the Gaussian similarity function (see Eq. 2.1), where the Euclidean distance is embedded, the parameter σ controls the width of neighborhood and the similarity is bound to $(0, 1]$. However, the cosine similarity metric (see Eq. 2.2) appears to be more effective for data in high-dimensional space [6]. **(2) How to construct the similarity graph.** There are several common ways, such as *full connection*, ε-*neighborhood* and *k-nearest neighbors* [10]. The first way generates a dense matrix. The last two ways yield typically a sparse similarity matrix by setting the similarity s_{ij} to zero if the distance between instances x_i and x_j is greater than a threshold (ε) or x_j is not among the nearest neighbors of x_i, respectively. However, the *k-nearest neighbors* seems more computationally expensive as it requires sorting operations.

Algorithm 1: Spectral clustering algorithm

Inputs: A set of data instances $X = \{x_1, ..., x_n\}$ with x_i in $\mathbb{R}^d$, the number of desired clusters k_c

Outputs: Cluster labels of n data instances

1 Construct similarity graph and generate similarity matrix S;
2 Derive unnormalized (L) or normalized (L_{sym} or L_{rw}) graph Laplacian;
3 Compute the first k_c eigenvectors of graph Laplacian, obtaining matrix U;
4 Normalize each row of matrix U to have unit length;
5 Perform k-means clustering on points defined by the rows of U;

$$\text{Gaussian similarity metric: } s_{ij} = \exp\left(-\frac{\|x_i - x_j\|_2^2}{2\sigma^2}\right) \qquad (2.1)$$

$$\text{Cosine similarity metric: } s_{ij} = \frac{x_i \bullet x_j}{\| x_i \| \| x_j \|} \qquad (2.2)$$

The generated similarity matrix S is symmetric and of $n \times n$ size. Then we derive the diagonal degree matrix D with $deg_i = \sum_{j=1}^n s_{ij}$. Next, we calculate the (unnormalized) graph Laplacian $L = D - S$ which does not depends on the diagonal elements of the similarity matrix and whose eigenvalues and eigenvectors are associated with many properties of graphs [10]. Moreover, L can be further normalized as the symmetric matrix $L_{sym} := D^{-1/2}LD^{-1/2}$ or the non-symmetric matrix $L_{rw} := D^{-1}L$. In order to achieve good clustering in broader cases, it is argued and advocated [10] to use normalized instead of unnormalized graph Laplacian, and in the two normalized cases to use L_{rw} instead of L_{sym}. Obviously, choosing a Laplacian matrix and its properties impacts the choice of solvers that can be used to calculate its eigenvectors (e.g. choosing L_{rw} will not allow the use of the `syevdx` symmetric eigensolver in the cuSOLVER library).

From the graph cut point of view, clustering on a dataset X is equivalent to partitioning a graph G into k_c partitions by finding a minimum balanced cut. *Ratio cut* and *normalized cut* are the two most common ways to measure the balanced cut, however minimizing *ratio cut* or *normalized cut* is an NP-hard optimization problem. Fortunately, the solution can be approximated from the first k_c eigenvectors (associated with the smallest k_c eigenvalues) of graph Laplacian matrix [10,11]. Let U denote the $n \times k_c$ matrix containing the eigenvectors

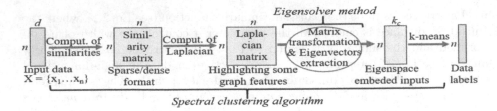

Fig. 2. Main computation steps in spectral clustering

as columns. Then each row of U can be regarded as the embedded representation in $\mathbb{R}^{k_c}$ of the original data instance in $\mathbb{R}^d$ with the same row number.

Finally, the k-means algorithm is applied on the embedded representation by regarding each row of the matrix U as a k_c-dimensional point, which therefore allows to find k_c clusters of original n data instances. In addition, before performing the final k-means, it is customary to scale each row of matrix U to unit length to improve the clustering result.

To summarize, spectral clustering involves several data transformation steps, illustrated in Fig. 2. A similarity matrix is computed based on the nature of the dataset and the clustering objective to model a connectivity graph, and then a Laplacian matrix is deduced, highlighting some information about the graph topology and the desired clustering. Eigenvectors are extracted, transcribing the information from the Laplacian matrix and allowing to form a $n \times k_c$ matrix where the n input data are encoded in the eigenspace of the first k_c eigenvectors. In this space, a simple k-means can then group the input data into k_c clusters.

3   Similarity Matrix Construction in CSR on GPU

In this section, we focus on the design of GPU parallel algorithms for constructing similarity matrix directly in CSR sparse format, in order to address both the $\mathcal{O}(n^2 d)$ computational challenge and the $\mathcal{O}(n^2)$ memory requirement of similarity matrix construction.

3.1   Need of Passing from Dense Format to Sparse Format

Initially we started from constructing the similarity matrix S simply in dense format on the GPU. Then we chose to compute the normalized graph Laplacian matrix L_{sym} (symmetric) instead of L_{rw} (non-symmetric) so that we could calculate its smallest k_c eigenvectors with syevdx, a dense symmetric eigensolver in the GPU-accelerated cuSOLVER library from NVIDIA. Finally, we applied our GPU parallel algorithm of k-means [5] on the points defined by the rows of k_c-eigenvector matrix to obtain the clustering result.

However, we quickly ran out of limited GPU memory when trying to process datasets with larger n since the similarity matrix and the graph Laplacian matrix were both constructed in dense format with $\mathcal{O}(n^2)$ memory requirement.

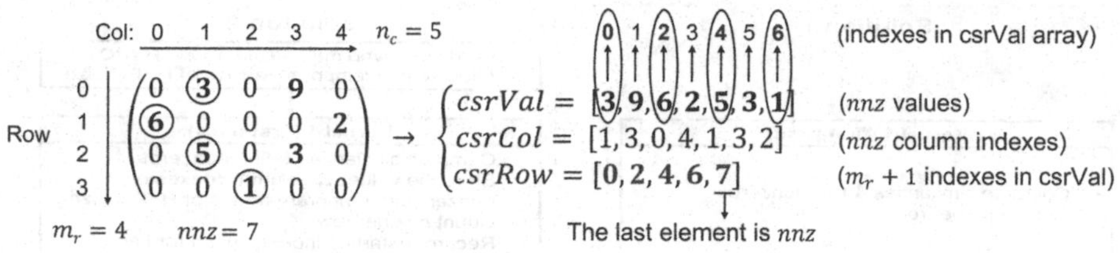

Fig. 3. An example for CSR format with an $m_r \times n_c$ matrix

On the other hand, the similarity matrix associated with ε-*neighborhood* graph or *k-nearest neighbors* graph generally has a sparse pattern, i.e. containing numerous zeros. Even for the similarity matrix associated with fully connected graph, we observed typically a significant portion of elements are close to zero. By setting a small threshold (e.g. 0.01) and regarding those under-threshold similarities as zeros, we are likely to obtain a sparse similarity matrix. Storing this array in a sparse format rather than a dense format significantly saves GPU memory, which greatly increases the scale of datasets that can be processed.

3.2 Choice of CSR Sparse Format

There are several commonly used sparse formats for storing a sparse matrix, such as Coordinate Format (COO), Compressed Sparse Row Format (CSR), Compressed Sparse Column Format (CSC), Ellpack, etc. In our case of sparse similarity matrix, we choose to use the CSR format for two reasons. First, the CSR format can usually achieve a good trade-off between memory space requirement and operation flexibility, and is efficient both for regular sparsity pattern and for power-law distribution [3]. With these advantages, the CSR format has been widely used and supported in most libraries. Second, we intend to employ the nvGRAPH spectral graph partitioning API but it supports only the CSR format for graph representation.

A sparse matrix represented in CSR format consists of three arrays. We call them `csrVal`, `csrCol`, `csrRow`. Figure 3 gives a simple example of CSR representation. The `csrVal` stores all nonzero values of the matrix in row-major format. The `csrCol` contains the column index of every nonzero element. Considering the first nonzero element in each row of the matrix (i.e. the circled red numbers in Fig. 3), the `csrRow` holds their indices that count in the `csrVal` array (i.e. the blue numbers circled by red ellipses) and contains in the end the total number of nonzeros in the matrix.

3.3 Difficulties

We have not found any existing work on the construction of similarity matrix in CSR format on the GPU, neither in the literature nor in numerous GPU-accelerated libraries. Studies [3,4,11,12] in the field of graph partitioning often

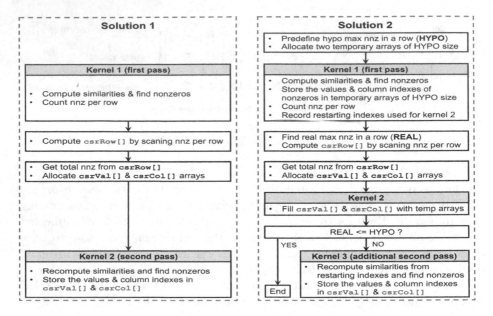

Fig. 4. Our two solutions for CSR similarity matrix construction on GPU

target graphs in COO topology represented by an edge list or in CSR topology represented by an adjacency list. They typically assume the availability of these sparse format lists, and do not consider or do not need the construction process. Another work [8] constructs sparse similarity matrix in COO format but on the assumption that the neighborhood information is given by an edge list.

In the standard case of spectral clustering, we have no such edge list or adjacency list available, but only n data instances in $\mathbb{R}^d$ dimensions. To obtain the similarity matrix in CSR format and save memory space, it makes no sense to first construct a sparse similarity matrix in dense format and then transform it from dense to CSR format. Thus, the construction of similarity matrix must be directly in CSR format. However, this is difficult to be done in parallel especially on the GPU, because (1) the total number of nonzeros is unknown, so we cannot allocate memory for the csrVal and csrCol arrays; (2) the number of nonzeros per row is unknown, so we cannot know in which segment of csrVal and csrCol we should store the value and column index of each nonzero, respectively; (3) although GPU threads can compute similarities and find nonzeros in parallel, they cannot parallelly store nonzeros (values and column indexes) at the right place of csrVal and csrCol, since each thread does not know the number of nonzeros ahead of it.

3.4 Our Solutions

We propose two solutions for the similarity matrix construction in CSR on GPU.

Table 1. Datasets and parameter settings of our benchmarks

Dataset	(n, d, k_c)	Similarity metric	Threshold	Block size of CUDA kernels
MNIST-60K	(60K, 784, 10)	Cosine	0.8 (similarity)	64
MNIST-120K	(120K, 784, 10)	Cosine	0.8 (similarity)	64
MNIST-240K	(240K, 784, 10)	Cosine	0.8 or 0.84^* (similarity)	64
Synthetic-1M	(1M, 4, 4)	Gaussian ($\sigma = 0.01$)	0.0002 (distance)	128
Synthetic-5M	(5M, 4, 4)	Gaussian ($\sigma = 0.01$)	0.0001 (distance)	128

Table 2. Performance comparison on GPU of our 2 solutions (S1 vs. S2)

Dataset	Max nnz in a row	Average nnz per row	Total nnz	Sparsity (% of 0)	S1 (s)	S2 (s) <1st HYPO>	<2nd HYPO>	Best speedup S2 vs S1
MNIST-60K	2196	251	15.1M	99.581%	7.16	5.29 <1024>	5.95 <2048>	**1.35**
MNIST-120K	3310	299	35.9M	99.751%	29.98	24.5 <1024>	24.39 <2048>	1.23
MNIST-240K	5552	478	114.8M	99.801%	126.85	103.75 <1024>	104.77 <2048>	1.22
MNIST-240K*	3520	199	47.8M	99.917%	125.17	91.75 <1024>	102.10 <2048>	**1.36***
Synthetic-1M	54	23	23.4M	99.998%	13.57	10.20 <16>	8.35 <54>	**1.63**
Synthetic-5M	64	29	149.9M	99.999%	362.79	318.45 <32>	312.48 <64>	1.16

Solution 1. As shown in the left part of Fig. 4, our Solution 1 mainly consists of two kernels with two complete passes across all the elements in similarity matrix. The first pass in Kernel 1 computes the similarities of all pair of instances, find the over-threshold similarities as nonzeros, and count the number of nonzeros (nnz) per row. Then the `csrRow` can be obtained by performing an exclusive scan on the array of nnz per row. Moreover, the total nnz can be known from the last element of `csrRow` and we can therefore allocate `csrVal` and `csrCol` arrays. With all these materials, we then launch the second pass in Kernel 2 to recompute all the similarities, find nonzeros as the first pass, and finally fill the `csrVal` and `csrCol` arrays.

Solution 2. As shown in the right part of Fig. 4, our Solution 2 is primarily composed of three kernels performing a single pass or possibly two passes (when running Kernel 3). It requires to predefine a hypothesis (HYPO) for the maximum number of nonzeros in a row, and allocate two temporary arrays of HYPO size for `csrVal` and `csrCol`. Then the first pass in Kernel 1 needs to undertake several tasks: not only compute all similarities, find nonzeros, count nnz per row, but also store the information of nonzeros in the temporary arrays, and meanwhile record the restarting places for the additional pass in Kernel 3 in case that our hypothesis is wrong. Then we can find the real maximal nnz in a row (REAL) from the nnz per row array. Next, we compute the `csrRow`, the total nnz, and allocate `csrVal` and `csrCol` arrays as our Solution 1. Thanks to the results stored in the temporary arrays in Kernel 1, we can just fill `csrVal` and `csrCol` arrays with Kernel 2. If our hypothesis is correct (REAL <= HYPO), then we will obtain the complete result in Kernel 2. Otherwise, we need to conduct an additional second pass in Kernel 3 to find the nonzeros out of hypothesis and store them at the right place in `csrVal` and `csrCol`. Particularly, the additional pass will not traverse all elements but will only start from the restarting indices recorded in Kernel 1.

Parallel Implementation on GPU. Both solutions are mainly implemented with our optimized CUDA kernels. For each of the CUDA kernels in Fig. 4, we create a 1D grid containing n 1D blocks. Each block of threads process one row of the matrix in a loop fashion. For Solution 2, we pay particular attention to the memory address alignment of restarting indexes on multiples of 32 memory words, which proves to have an significant impact on performance.

3.5   Experiments with Our Two Solutions

We tested our two solutions on a GeForce RTX 3090 with the datasets and parameter settings shown in Table 1. For the reason of comparison, we set the same threshold (0.8) for all MNIST-based datasets (https://leon.bottou.org/projects/infimnist). However, only the first two datasets yield satisfying clusterings with this threshold, while the MNIST-240K needs a higher threshold (0.84, marked with *) to generate a good clustering. Before clustering, we perform feature scaling for the synthetic datasets to transform all values into the same scale within the range [0,1]. Note that we do not have a version or benchmark on CPU yet.

Table 2 shows the results of our benchmarks. We observed that the similarity matrices are extremely sparse although they contain millions of nonzeros. Both two solutions are scalable up to 5 million instances. With appropriate hypotheses, our Solution 2 outperforms Solution 1 with a speedup from ×1.16 up to ×1.63.

However, this is conditional. Figure 5 shows the impact of hypothesis on the performance of our Solution 2 with the MNIST-120K set. When the hypothesis is small (HYPO < 256) or very large (HYPO > 2399), our Solution 2 is less efficient than Solution 1. Specifically, a smaller hypothesis will leave more computations and more memory accesses to Kernel 3 of Solution 2, while a greater hypothesis will need Kernel 1 of Solution 2 to compute more and record more in global memory. In particular, we observed two sudden increases of time when the hypothesis grows from 2398 to 2399 and from 3038 to 3039. We think they are caused by the decrease of the number of resident blocks in Stream Multiprocessors, because we use much shared memory when the hypothesis is large. Although our Solution 2 can be influenced by the value of hypothesis, we found a fairly wide range [256, 2399] where Solution 2 achieves better performance than Solution 1. Hence finding an appropriate hypothesis should not be difficult if we avoid extreme values. The Kernel 2 of Solution 2 consumes little time compared to other kernels, so is omitted here.

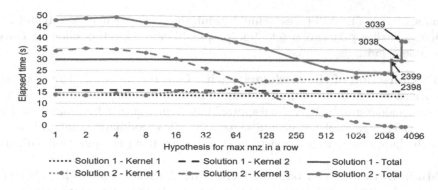

Fig. 5. Detailed comparison of our 2 solutions on the MNIST-120K dataset

4   Eigenvector Extraction for Spectral Clustering on GPU

4.1   Eigensolver Methods

As presented in Sect. 2, a key step of spectral clustering is to calculate the first k_c eigenvectors of Laplacian matrix. This can be done with eigensolver methods. They include new matrix transformations to facilitate the eigenvectors extraction and are not specific to spectral clustering. Three well-known methods are the following:

- **Arnoldi's** method [13]: it takes any input matrix (like L, L_{sym} or L_{rw}, see Sect. 2) and transforms it into an Hessenberg matrix, then calls an eigensolver (usually the solver QR). This is a generic but computationally expensive method.
- **Lanczos** method [13]: similar to Arnoldi's method but requires a real and symmetric (or Hermitian) input matrix (like L or L_{sym}) which it transforms into a tridiagonal matrix, before calling an eigensolver (like QR). This is an efficient method but it suffers from numerical instabilities.
- **LOBPCG** method [9]: requires a symmetric input matrix (like L or L_{sym}) or a pair of matrices with one symmetric and one symmetric positive definite (like (L, D)), then starts extracting the smallest k_c eigenpairs. The LOBPCG method performs some transformations of the matrices and calls other eigensolvers on smaller internal submatrices. LOBPCG is more recent (released in 2000) than the previous two methods.

Implementations of these algorithms exist in different libraries. They require input matrices in dense or sparse format and are sometimes improved to be more robust to numerical instabilities.

4.2   Eigensolver-Embedded Algorithms in the nvGRAPH Library

With the similarity matrix in CSR sparse format defined in Sect. 3, the remaining steps of spectral clustering can be completed by calling the spectral graph partitioning API of nvGRAPH library [12], designed by NVIDIA for GPU-accelerated graph analytics. The API takes as input graph the similarity matrix only in CSR format, and then performs spectral graph partitioning with the following three selectable algorithms:

- **Maximization of the *modularity* with Lanczos method.** The *modularity* measures how well a partitioning applies to the target graph compared to a random graph. It can be approximately maximized by looking at the largest eigenpairs of *modularity* matrix [3]. This modularity matrix instead of the Laplacian matrix is constructed before eigenpair computation with the Lanczos solver.
- **Minimization of the balanced cut with Lanczos method.** This algorithm aims at minimizing the volume of inter-cluster connections relative to the size of clusters (i.e. balanced cut). It constructs the Laplacian matrix and then calls the Lanczos solver.
- **Minimization of the balanced cut with LOBPCG method.** Similar principle to the second algorithm, but utilizes the LOBPCG solver to handle the constructed Laplacian matrix.

Compared to Lanczos method, LOBPCG can handle eigenvalues with multiplicity (which often happens in spectral clustering). Moreover, the NVIDIA implementation is able to restart the computation when it encounters numerical instabilities. Thus this LOBPCG-embedded algorithm has appeared as the most reliable solution on our benchmarks. Note that all the three algorithms contain the k-means step at the end.

4.3   Experiments with nvGRAPH Library

Following the similarity matrix construction in CSR format (Sect. 3), we tested the nvGRAPH spectral graph partitioning API to complete spectral clustering on the GPU. The LOBPCG-embedded algorithm is selected for our benchmarks. Several parameters need to be specified. We simply set the maximal number of iterations to the nvGRAPH default value, i.e. 4000 for the LOBPCG eigensolver and 200 for the final k-means. However, we found that the approximation tolerance for the eigensolver needs to be tuned because it has a significant impact on the clustering quality and the execution time. Depending on the benchmarks, a too small tolerance may lead to eigensolver divergence and too much execution time, while a too large tolerance can result in bad clusterings. Besides, the tolerance for k-means is set to 0.0001.

Table 3 presents the elapsed time of our spectral clustering (including the nvGRAPH API) on the GeForce RTX 3090, as well as the clustering quality measured by three metrics: Rand Index (RI), Ajusted Rand Index (ARI), Normalized Mutual Information (NMI) [15]. The last two metrics are stricter than the first one. All metrics return a score less or equal to 1, and a score closer to 1 indicates a better clustering. Our benchmarks on the MNIST-based datasets (handwritten digits) yielded relatively good clustering quality, while we found perfect clusterings on the synthetic datasets (convex clusters).

Table 3. Elapsed time and clustering quality of spectral clustering on GPU

Dataset	Elapsed time (s)					Quality metric		
	Data transfers	Similarity matrix constr. in CSR	nvGRAPH LOBPCG <eigen. tolerance>		Total	RI	ARI	NMI
MNIST-60K	0.015	5.29	2.35 <0.005>		7.66	0.93	0.63	0.74
MNIST-120K	0.031	24.39	4.03 <0.005>		28.45	0.89	0.50	0.65
MNIST-240K	0.062	91.75	5.42 <0.005>		97.23	0.88	0.47	0.69
Synthetic-1M	0.002	8.35	3.61 <0.001>		11.96	1.00	1.00	1.00
Synthetic-5M	0.008	312.48	29.75 <0.0001>		342.24	1.00	1.00	1.00

With respect to the elapsed time, the similarity matrix construction turns out to be the most time-consuming step of spectral clustering, mainly due to its $\mathcal{O}(n^2 d)$ time complexity. Although the theoretical complexity of eigenvector computation is $\mathcal{O}(n^3)$, the nvGRAPH LOBPCG eigensolver did not take much time compared to the similarity matrix construction. This is mainly due to the fact that the eigensolver adopts an iterative and approximate method instead of direct methods. The data transfers between CPU to GPU appears negligible. Finally, the large datasets demonstrate the scalability on a GPU of our implementation of spectral clustering with algorithms optimized for CSR sparse matrix generation (although it still dominates the runtime).

5 Conclusion and Future Work

We have presented our scalable parallel algorithms for spectral clustering on a single GPU. We have proposed two solutions for the construction of similarity matrix directly in CSR sparse format on GPU, which can save a large amount of GPU memory space compared to dense format storage. Moreover, our matrix generation in CSR format is compatible with nvGRAPH's eigensolver-embedded algorithms that require CSR matrices. With our sparse matrix construction and nvGRAPH's eigensolvers, we have obtained a parallelized end-to-end spectral clustering implementation on one GPU. Finally, our experiments show that our GPU implementation succeeds to scale up to millions of data instances.

To address even larger datasets, it would be interesting to parallelize spectral clustering on multi-GPU machines which provide more computing power and memory space. Another solution would be CPU-GPU algorithms incorporating the *representative* extraction technique on the CPU to reduce the number of data instances to process on the GPU. Moreover, a comparison with a purely CPU multithreaded and vectorized version would be interesting in the future.

Acknowledgement. This work was supported in part by the China Scholarship Council (No. 201807000143). The experiments were conducted on the research computing platform supported in part by Région Grand-Est, Metz-Métropole and Moselle Departement.

References

1. Arthur, D., Vassilvitskii, S.: k-means++: the advantages of careful seeding. In: Proceedings of the Eighteenth Annual ACM-SIAM Symposium on Discrete Algorithms, New Orleans, Louisiana, USA (2007)
2. Domingos, P.: A few useful things to know about machine learning. Commun. ACM **55**(10), 78 87 (2012)
3. Fender, A.: Parallel solutions for large-scale eigenvalue problems arising in graph analytics. Ph.D. thesis, Université Paris-Saclay (2017)
4. Fender, A., Emad, N., et al.: Accelerated hybrid approach for spectral problems arising in graph analytics. Procedia Comput. Sci. **80**, 2338–2347 (2016)
5. He, G., Vialle, S., Baboulin, M.: Parallel and accurate k-means algorithm on CPU-GPU architectures for spectral clustering. Concurr. Comput. Pract. Exp., e6621 (2021). https://onlinelibrary.wiley.com/doi/abs/10.1002/cpe.6621
6. Ina, T., Hashimoto, A., Iiyama, M., Kasahara, H., Mori, M., Minoh, M.: Outlier cluster formation in spectral clustering. arXiv preprint arXiv:1703.01028 (2017)
7. Jain, A.K.: Data clustering: 50 years beyond k-means. Pattern Recogn. Lett. **31**(8), 651–666 (2010)
8. Jin, Y., JáJá, J.F.: A high performance implementation of spectral clustering on CPU-GPU platforms. In: 2016 IEEE International Parallel and Distributed Processing Symposium Workshops, Chicago, IL, USA, pp. 825–834 (2016)
9. Knyazev, A.V.: Toward the optimal preconditioned eigensolver: locally optimal block preconditioned conjugate gradient method. SIAM J. Sci. Comput. **23**(2), 517–541 (2001)
10. von Luxburg, U.: A tutorial on spectral clustering. Stat. Comput. **17**(4), 395–416 (2007)
11. Naumov, M., Moon, T.: Parallel spectral graph partitioning. Technical report, NVIDIA Technical Report, NVR-2016-001 (2016)
12. NVIDIA: NVGRAPH library user's guide (2019)
13. Saad, Y.: Numerical Methods for Large Eigenvalue Problems. SIAM, Philadelphia (2011)
14. Sundaram, N., Keutzer, K.: Long term video segmentation through pixel level spectral clustering on GPUs. In: IEEE International Conference on Computer Vision Workshops, ICCV 2011 Workshops, Barcelona, Spain (2011)
15. Vinh, N.X., Epps, J., Bailey, J.: Information theoretic measures for clusterings comparison: variants, properties, normalization and correction for chance. J. Mach. Learn. Res. **11**, 2837–2854 (2010)
16. Xiang, T., Gong, S.: Spectral clustering with eigenvector selection. Pattern Recognit. **41**(3), 1012–1029 (2008)
17. Yan, D., Huang, L., Jordan, M.I.: Fast approximate spectral clustering. In: Proceedings of the 15th ACM International Conference on Knowledge Discovery and Data Mining, Paris, France, 2009 (2009)
18. Zelnik-Manor, L., Perona, P.: Self-tuning spectral clustering. In: Advances in Neural Information Processing Systems 17 (NIPS 2004), Vancouver, Canada, 13–18 December 2004, pp. 1601–1608 (2004)
19. Zheng, J., Chen, W., Chen, Y., Zhang, Y., Zhao, Y., Zheng, W.: Parallelization of spectral clustering algorithm on multi-core processors and GPGPU. In: 2008 13th Asia-Pacific Computer Systems Architecture Conference, pp. 1–8. IEEE (2008)

XSP: Fast SSSP Based on Communication-Computation Collaboration

Xinbiao Gan, Wen Tan, Menghan Jia, Jie Liu, and Yiming Zhang[✉]

School of Computer, NUDT, Changsha, China
{xinbiaogan, jiamenghan12, liujie}@nudt.edu.cn,
zym@nicexlab.com

Abstract. Single-source shortest path (SSSP) is an important graph search algorithm for data-intensive applications which finds the minimum distance from a source vertex to any other vertex in a given graph. Although having been extensively studied for both single- and multi-node scenarios, SSSP search still brings severe challenge to communication when processing large graphs that consist of billions of vertices involving hundreds of computing nodes. To address this problem, in this paper we propose XSP, a fast SSSP search method based on communication-computation collaboration, which optimizes the communication of parallel SSSP in two aspects. First, we design a group-based scalable batching mechanism which effectively reduces the inter-machine communication overhead. Second, we propose a CCO (Communication-Computation Overlapping) method which realizes non-blocking execution of communication and computation. We have implemented XSP and extensive evaluation results show that the performance of XSP is significantly higher than that of the state-of-the-art parallel SSSP methods.

Keywords: Single-source shortest path (SSSP) · XSP · Communication reduction · Message batching · Non-blocking execution

1 Introduction

Large-scale graphs have been applied to many practical scenarios in the world. Graphs play an important role in big data analysis such as social interaction, email networks, website links, etc. In these scenarios, graphs are used to describe the relationship (i.e., edges) between entities (i.e., vertices). With the explosion of data, very large graphs have recently emerged in various application fields. These applications require fast and scalable graph analysis, and consequently distributed graph processing has been extensively studied.

Single-source shortest path (SSSP) is an important graph search algorithm for data-intensive applications, which aims to find the minimum distance from a source vertex to any other vertex in a given graph. SSSP has been widely used in graph data analysis including social networks, business intelligence, and public safety. Massive graph search and traversal like SSSP and BFS (breadth-first search) have typical characteristics such as poor data locality and irregular memory access. To address these challenges, researchers have conducted a considerable amount of works on efficient

C. Cérin et al. (Eds.): NPC 2021, LNCS 13152, pp. 53–63, 2022.
https://doi.org/10.1007/978-3-030-93571-9_5

communication and computation of graph processing on distributed parallel systems for improving graph traversal performance [1–6].

With such high interest in analytics of large graphs, the Graph500 benchmark was proposed and is becoming more and more popular in recent years. Graph500 measures data analytics performance on data-intensive applications with the metric TEPS (Traversed Edges Per Second). Graph500 is particularly suitable for evaluating graph traversal, and currently Graph500 mainly has two tests, namely, the SSSP benchmark and the BFS benchmark.

The key characteristic of distributed graph traversal (like SSSP) is that it brings huge amount of inter-machine communication overhead. In most cases, the overhead of inter-machine communication is much higher than that of computation. To exploit the data processing potential and the powerful capacity of large-scale systems, recent studies propose to parallelize graph traversal using MPI (Message Passing Interface), a language-independent communication protocol that coordinates the computing tasks in parallel programs. However, MPI communication optimization requires a deep understanding of the characteristics of the applications running on large-scale systems [7], and thus is rather difficult for graph application developers. As a result, although having been extensively studied for both single- and multi-machine scenarios, SSSP search still brings severe challenge to communication when processing large graphs that consist of billions of vertices involving hundreds of machines.

To address this problem, in this paper we propose XSP, an extremely-fast SSSP search method based on communication-computation collaboration, which optimizes the communication of parallel SSSP in two aspects. First, we design a group-based scalable batching mechanism which effectively reduces the inter-machine communication overhead. Second, we propose a CCO (Communication-Computation Overlapping) method which realizes non-blocking execution of communication and computation.

We have implemented XSP. Extensive evaluation results show that the performance of XSP is significantly higher than that of the state-of-the-art parallel SSSP methods. We have also deployed XSP on the Tianhe Exascale Supercomputer [8] (2048-node Prototype) and ranked No.1 in the latest (June 2021) Graph500 list [9].

2   Background and Related Work

2.1   Single-Source Shortest Path

Single-Source Shortest Path (SSSP) tries to find the minimum distance from a source vertex to any other vertex in a given graph. Specifically, given a directed graph $G = (V, E)$, with non-negative costs on each edge, and a selected source node v in V, for all w in V, find the cost of the least cost path from v to w. The cost of a path is simply the sum of the costs on the edges traversed by the path. This problem is a general case of the more common subproblem, in which we seek the least cost path from v to a particular w in V. In the general case, this subproblem is no easier to solve than the SSSP problem.

2.2 Parallel Graph Processing

Parallel processing faces difficulty in graph partitioning, which is crucial for both reducing the communication cost and balancing the load in graph-parallel systems [10]. Early graph-parallel systems partition a graph by cutting the edges to evenly distribute the vertices. These edge-cut systems suffer from imbalanced computation and communication for high-degree vertices. In contrast, vertex-cut systems cut the vertices and evenly distribute the edges among machines. Although vertex-cut alleviates the imbalance problem of high-degree vertices, it incurs high communication cost and excessive memory consumption for low-degree vertices.

Essentially, the partitioning procedure divides a graph into n sub-graphs in an n-machine cluster, each being assigned to one worker machine. TopoX [23] proposes the topology refactorization strategy to address this problem. It refactorizes the graph with fusion and fission when storing graph vertices and edges. Specifically, fusion organizes a set of neighboring low-degree vertices into a super-vertex (for processing locality), and fission splits a high-degree vertex into a set of sibling sub-vertices (for load balancing). Refactorization is performed by all workers on CNs in parallel and we resolve the potential conflict by conducting double-check on vertex states. A worker checks whether the vertex has been processed both before moving it to the processing queue and before performing fusion and fission operations, and if a conflict occurs further processing will be skipped. This paper will adopt the refactorization strategy of TopoX [23] for graph partitioning, and focus on reducing the communication overhead between computing nodes during graph processing of SSSP.

2.3 Communication Optimization for Graph Processing

Communication performance is of paramount im-portance to parallel-computing applications. Today, MPI (Message Passing Interface) is the predominant parallel programming model for high-performance parallel computing, making it a key technology to be optimized so that scientific computing applications can take full advantage of the parallel computing system that they use. Low-level optimization requires a detailed understanding of the usage characteristics of applications on production parallel computing systems. Unfortunately, the performance of MPI implementations on large-scale clusters is significantly impacted by factors including its inherent buffering, type checking, and other control overheads. Researchers have done a considerable amount of works on communication optimizations [9–11].

In order to well understand MPI usage characteristics, researchers design Autoperf to gather insights from detailed analysis of the MPI logs, revealing that large-scale applications running on clusters tend to require more communication and parallelism. While MPI library does not behavior well and performs discrepancy vary from clusters from different vendors. Accordingly, M. Blocksome et al. designed and implemented a one-sided communication interface for the IBM Blue Gene/L supercomputer [12], which improved the maximum bandwidth by a factor of three. Furthermore, Sameer Kumar et al. presented several optimizations by extensively exploiting IBM Blue Gene/P interconnection networks and hardware features and enhancements to achieve near-peak performance across many collectives for MPI collective communication on Blue Gene/P [13].

For better support of task mapping for Blue Gene/L supercomputer, a topology mapping library is used in BG/L MPI library for improving communication performance and scalability of applications [14]. However, comprehensive topology mapping library might benefit by providing scalable support of MPI virtual topology interface. Moreover, Kumar et.al presented PAMI (Parallel Active Message Interface) as Blue Gene/Q communication library solution to the many challenges with unprecedented massive parallelism and scale [15]. In order to optimize the performance of large-scale applications, IBM developed LAPI (Low-level Ap-plications Programming Interface), a low-level, high-performance communication interface available on the IBM RS/6000 SP system [16]. It provides an active message-like interface along with remote memory copy and synchronization functionality. However, the limited set from LAPI does not compromise on functionality expected on a communication API, whats worse is that topology mapping library and LAPI is designed for IBM series supercomputers, resulting in difficulties in adaptation to applications running on general clusters.

Different from communication optimizations on IBM series supercomputers, Naoyuki Shida et al. implemented a customized MPI library and low-level communication at tofu topology level based on Open MPI for K computer [17]. Above proposed MPI implementation is target at K computer. Similar problems could be found in communication related studies for supercomputers [17–22].

Anton Korzh from Graph500 executive committee designed AML (Active Messages Library) which has been included in the Graph500 reference code [9]. AML is an SPMD (Single Program Multiple Data) communication library built on top of MPI3 intended to be used in fine grain applications like Graph500. AML would make user code clarity while delivering high performance, However, current version of AML support only one-sided message, which cannot send a response from active message handler and there are not two-sided active messages, which could facilitate graph traversal like SSSP.

3 XSP Design

3.1 Overview

In this section, we will introduce our optimization of Single-Source Shortest Path (SSSP) algorithm. Graph traversal algorithms are often evaluated by using the TEPS (Traversed edges per second) metric, which is proposed by the Graph500 benchmark. Different from FLOPS (Floating Point Per Second) for compute-intensive workloads proposed by Top500 Linpack benchmark [17], TEPS is defined as

$$TEPS = edges\_traversed/SSSP\_time \tag{1}$$

Given a graph, the number of traversed edges is a constant, and thus we can improve the TEPS performance of SSSP only by reducing the SSSP time. Previous studies on graph traversal algorithms have shown that the communication bottleneck is mainly because of the huge numbers of small messages when traversing the graphs, which motivates us to design XSP, which improves the SSSP performance by combining the huge numbers of small messages into batched large messages.

In the rest of this section, Sect. 3.2 will introduce the compact storage mechanism of XSP, Sect. 3.3 will introduce the group-based scalable batching mechanism which effectively reduces the inter-machine communication overhead, and Sect. 3.4 will propose a CCO (Communication-Computation Overlapping) method which realizes non-blocking execution of the communication and computation of graph traversal.

3.2 Graph Storage

In XSP, the graph data is stored as the *adjacency matrix* and converted into a compressed sparse row (CSR) matrix, the process of which is as follows.

Suppose that the adjacency matrix is of $n * n$, and the number of stored edges is m. There are three arrays in the CSR matrix, namely, the *row offset* array (of length $n + 1$), the *column indices* array (of length m), and the *values* array (of length m).

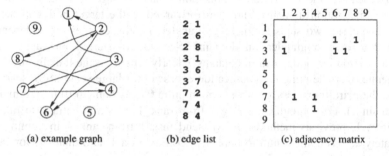

(a) example graph   (b) edge list   (c) adjacency matrix

Fig. 1. Adjacency matrix example with nine vertices and nine edges.

For a directed graph, the number of edges stored in the ith row of the *adjacency matrix* is the $(i + 1)^{th}$ element of *row offset* array subtracts the ith element indicate, and the *column indices* array stores the indices of the target vertices of the edges in the *adjacency matrix* row by row. The *values* array records the values in the adjacency matrix.

For example, for Fig. 1, the row offset array is: 0, 0, 3, 6, 6, 6, 6, 8, 9, 9, representing there are three, three, two, and one edges in the 2nd, 3rd, 7th, and 8th rows, respectively. And the column indices array is: 1, 6, 8, 1, 6, 7, 2, 4, 4, respectively represents the indices of the edges in the rows, namely, (1, 6, 8) for row 2, (1, 6, 7) for row 3, (2, 4) for row 7, and (4) for row 8. The values array simply includes nine "1"s. For SSSP, the *values* array can be ignored if the edges have no weights. Therefore, only $n + m + 1$ numbers are needed to represent all the edges, in comparison with the $2m$ numbers needed by the original adjacency matrix.

3.3   Group-Based Hierarchical Batching

In large-scale clusters, the global communication could be refined into hierarchical groups. Taking an example, a three-level fat-tree, where the level-0 communication is within the lowest switches (e.g., 0000−0001) and the level-2 communication is across the highest switches (e.g., 0000−1111). Logically, the processes with the same IP are painted with the same color as they arc on the same node. The processes in the global communication are divided into hierarchical communication domains. Processes with the same local position in all domains of level i are classified into an inter-domain communication group for level i.

In most cases a two-level cluster can accommodate hundreds of nodes, which is enough for processing SSSP for billions of vertices. Therefore, for simplicity we will consider tow-level clusters in the rest of this paper (but it is natural for our design to extend to multi-level clusters). If all processes are viewed as a matrix, then the processes of the low-level domain will form a column, and the processes with the same local ID of each low-level domain will form a row.

Previous graph-parallel systems perform SSSP search without taking communication overhead as the first-order optimization goal. Messages are firstly transferred across communication domains, and then forwarded to the target nodes in the same domain. There are two steps for message transfer. First, messages are independently transferred across communication domains. Second, transferred messages are forwarded to destination nodes within domains. Clearly, this message transfer mechanism will generate overwhelming communication overhead for large-scale SSSP search. For example, there are four messages from comm_intra 0 (i.e., domain 0) to comm_intra 1 (i.e., domain 1). For example, for msg 1 from rank_1 in comm_intra 0 to rank_2 in comm_intra 1, previous methods will (i) send msg 1 from rank_1 in comm_intra 0 immediately across comm_intra to rank_1 in comm_intra 1, and then (ii) forward the message to rank_2 in comm_intra 1.

In contrast, XSP proposes to hierarchically group the messages of parallel SSSP into batches. Messages should be first gathered in communication domain, and then forwarded (as a big message) to each destination domain. Compare the state-of-the-art SSSP methods and XSP, the main difference is the message transfer flows. In previous studies, sending message will first cross communication domain and then forwarded within each domain, while in XSP sending message will be gathered in communication domains before being forwarded to other domains. Therefore, XSP takes communication overhead as the first-order optimization goal and thus has lower communication overhead than before.

XSP realizes message batching of parallel SSSP as follows. (i) A relay node R_1 gathers messages in the communication domain and pack into a long message. (ii) R_1 forwards the long message to the target domain (across different domains). (iii) The corresponding relay node R_2 in the target domain scatters the packed message to the destination nodes. Suppose that there are four messages from comm_intra 0 (i.e., domain 0) to comm_intra 1 (i.e., domain 1). Firstly, the four messages are gathered into rank_3 in comm_intra 0 and then packed into a long message. Secondly, the long message would be forwarded to rank_3 comm_intra 1 across domains. Thirdly, Scatter the packed message to destination nodes in comm_intra 1. Clearly, the batching mechanism of XSP substantially reduces the number of messages and thus effectively improves the communication efficiency.

3.4   Communication-Computation Overlapping

We propose a CCO (Communication-Computation Overlapping) mechanism for XSP to realize the non-blocking execution of the communication and computation of SSSP search, as illustrated in the following Algorithm 1.

Algorithm 1. The pseudo code of CCO

```
1.  init()
2.  //custom function by user for comm_inter
3.  void xsp_register_handler(void(*f)(int, void*, int),int n)
4.  // there are three parameters in f(int, void*, int), rank for source process, message
pointer and len respectively
5.  dividing communication into comm_intra and comm_inter
6.  create buffers for receiving server and the sending server, respectively
7.  voting route process in comm_intra
8.  define messages msgs.
9.  listening msgs from both comm_intra and comm_inter at route
10.   if route.msg->rank belongs to comm_intra.route //msg from same comm_intra
11.   then
12.       comm_intra.buffer[i]←msgs
13.   // according to the target process of data trans-mission
14.           if  comm_intra.buffer[i] ->size ≥thr
15.           //thr is defined by ueser
16.           then
17.                   gather scattered msgs to route
13.           endif
14.   else
15.       route.msg ->rank comm_inter.route
16.       //msg from across comm_intra
17.   endif
18.   call xsp_register_handler(void(*f)(int, void*, int),int n)
19.   barrier()
20.   if all msgs from intra and inter to route is done
21.   then
22.       exit
23.   else goto 8
24.   end if
```

Specifically, CCO has the following specific designs.

(1) The receiving server and the sending server would receive and transmit data asynchronously using repeated non-blocking communication respectively to realize the overlapping of calculation and communication.

(2) The communication flow in XSP must gather and pack scattered messages firstly within the communication domain, then forward message across domains, and finally scatter packed messages into destination in communication domains for reducing the communication traffic.

(3) Merge and send messages according to target process with four buffers by default to further reduce communication overhead.

(4) XSP uses the standard MPI interface to transfer and handle different types and sizes of message, so as to reduce the difficulty of programming and modification.

4   Evaluation

We have implemented XSP. Extensive evaluation results show that the performance of XSP is significantly higher than that of the state-of-the-art parallel SSSP methods.

Table 1. Testbed configurations.

Parameter	Configuration
Num. of Nodes	128
Network	2-level tree (200 Gbps)
oversubscription	1:4
CPU	2 GHz
Cores/Node	384
Total num. of Cores	196608
Memory/Node	192 GB
Memory (GB)	98304

4.1   Methodology

We use the Graph500 SSSP benchmark to evaluate performs SSSP to a Kronecker graph modeling real-world networks from selected 64 roots randomly. Graph500 benchmark performs the following steps according to specification and reference implementation [8].

In order to validate the proposed XSP method for SSSP search, we conduct extensive experiments on a 128-node testbed, as listed in Table 1.

The 128-node network is organized as a two-level tree structure. In our evaluation, the graph scale increases from 26 to 30, with the number of vertices increases from 64 million to 1 billion. In all experiments we adopt one-sided messages for simplicity, and each result is the mean of 20 runs without specification.

4.2   SSSP Performance

We evaluate XSP-based SSSP and compare it with the original implementation of SSSP provided by Graph500. As illustrated in Fig. 2. XSP-based SSSP performs much better than that of the original SSSP. This is mainly because XSP greatly reduces the communication overhead and thus significantly boost the overall SSSP performance.

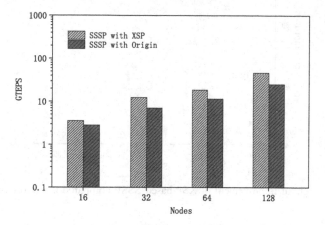

Fig. 2. SSSP performance comparison.

4.3 Communication Time

In order to validate the effectiveness of the communication reduction design of XSP, we conduct communication performance comparison and analysis on XSP-based SSSP and the original SSSP. Table 2 compares the communication times of XSP-based SSSP and the original SSSP.

From Table 2, we can find that as scale increases from 26 to 30, the communication time of XSP-based SSSP is much less than that of the original SSSP. It is obvious that XSP performs much better than previous designs in communication reduction, which is mainly owing to its hierarchical message batching mechanism.

Table 2. Communication time comparison.

Scale	Configuration time (seconds)	
	Origin	XSP
26	1.592161021	1.078084771
27	3.307811271	2.080718521
28	7.138187167	4.432171979
29	15.55900481	10.70678908
30	31.95777698	28.14610063

4.4 Communication Throughput

We further compare the communication throughput of XSP-based SSSP and the original SSSP, where the throughput is the transferred data volume per minute. The communication throughput comparison between XSP-based SSSP and the original SSSP is demonstrated in Fig. 3.

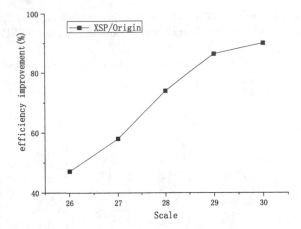

Fig. 3. Throughput improvement of XSP-based SSSP compared to the original SSSP.

As demonstrated in Fig. 3, XSP-based SSSP is rather better than the original SSSP, especially by up to more than 80% at scale = 30. The efficiency improvement becomes slightly slow from scale = 29. The main reason is that immediately forwarding is not conductive to the full utilization of networking bandwidth. Based on the result of Fig. 3, we can see that XSP-based SSSP has overwhelming advantage over the original SSSP, no matter how large the graph is.

5 Conclusion

In this paper we propose XSP, an extremely-fast SSSP method which optimizes the communication of parallel SSSP in two aspects. First, we design a group-based scalable batching mechanism which effectively reduces the inter-machine communication overhead. Second, we propose a CCO (Communication-Computation Overlapping) method which realizes non-blocking execution of communication and computation. We have implemented a prototype of XSP. Extensive evaluation results show that the performance of XSP is significantly higher than that of the state-of-the-art parallel SSSP methods. In our future work, we will apply the key idea of XSP to other graph traversal algorithms like BFS (Breadth-First Search). We will also study the use of two-sided messages for XSP-based SSSP.

References

1. Eason, G., Noble, B., Sneddon, I.N.: On certain integrals of Lipschitz-Hankel type involving products of Bessel functions. Phil. Trans. Roy. Soc. London **A247**, 529–551 (1955)
2. Ueno, K., Suzumura, T.: Highly scalable graph search for the Graph500 benchmark. In: Proceedings of the 21st International Symposium on High-Performance Parallel and Distributed Computing, New York, NY, USA (2012)

3. Fuentes, P., Bosque, J.L., Beivide, R., Valero, M., Minkenberg, C.: Characterizing the communication demands of the Graph500 benchmark on a commodity cluster. In: Proceeding of the 2014 IEEE/ACM International Symposium on Big Data Computing, London, UK (2014)
4. Suzumura, T., Ueno, K., Sato, H., Fujisawa, K., Matsuoka, S.: Performance characteristics of Graph500 on large-scale distributed environment. In: Proceeding of the 2011 IEEE International Symposium on Workload Characterization (IISWC), Austin, TX, USA (2011)
5. Nakao, M., Ueno, K., Fujisawa, K., Kodama, Y., Satoh, M.: Performance evaluation of supercomputer Fugaku using breadth-first search benchmark in Graph500. In: Proceeding of the 2020 IEEE International Conference on Cluster Computing (CLUSTER), Kobe, Japan (2020)
6. Ueno, K., Suzumura, T.: Book 2D Partitioning Based Graph Search for the Graph500 Benchmark, pp. 1925–1931. IEEE (2012)
7. Beamer, S., Asanovic, K., Patterson, D.: Book Direction-optimizing Breadth-First Search, pp. 1–10. IEEE (2012)
8. Wang, R., et al.: Brief introduction of TianHe exascale prototype system. Tsinghua Sci. Technol. **26**(3), 361–369 (2021)
9. Graph500 https://www.graph500.org/
10. Li, S., Huang, P.-C., Jacob, B.: Exascale Interconnect Topology Characterization and Parameter Exploration, pp. 810–819. IEEE (2018)
11. Zhu, Y., Taylor, M., Baden, S.B., Cheng, C.-K.: Advancing Supercomputer Performance Through Interconnection Topology Synthesis, pp. 555–558. IEEE (2008)
12. Blocksome, M., et al.: Design and Implementation of a One-Sided Communication Interface for the IBM eServer Blue Gene, p. 54. IEEE (2006)
13. Faraj, A., Kumar, S., Smith, B., Mamidala, A., Gunnels, J.: MPI Collective Communications on The Blue Gene/P Supercomputer Algorithms and Optimizations, pp. 63–72. IEEE (2009)
14. Yu, H., Chung, I.-H., Moreira, J.: Topology Mapping for Blue Gene/L Supercomputer, p. 52. IEEE (2006)
15. Kumar, S., et al.: PAMI A Parallel Active Message Interface for the Blue Gene/Q Supercomputer, pp. 763–773. IEEE (2012)
16. Shah, G., et al.: Performance and Experience with LAPI-a New High-Performance Communication Library for the IBM RS/6000 SP, pp. 260–266. IEEE (1998)
17. Shida, N., Sumimoto, S., Uno, A.: MPI library and low-level communication on the K computer. Fujitsu Sci. Tech. J. **48**(3), 324–330 (2012)
18. Li, M., Lu, X., Potluri, S., Hamidouche, K., Tomko, J.J.K., Panda, D.K.: Scalable Graph500 design with MPI-3 RMA, pp. 230–238. IEEE (2014)
19. Dijkstra, E.W.: A note on two problems in connection with graphs. Numer. Math. **1**(1), 269–271 (1959)
20. Nikas, K., Anastopoulos, N., Goumas, G., Koziris, N.: Employing Transactional Memory and Helper Threads to Speedup Dijkstras Algorithm, pp. 388–395. IEEE (2009)
21. Bellman, R.: On a routing problem. Q. Appl. Math. **16**, 88–90 (1958)
22. Plimpton, S.J., Devine, K.D.: MAPREDUCE in MPI for large-scale graph algorithms. Parallel Comput. **37**(9), 610–632 (2011)
23. Zhang, Y., et al.: TopoX topology refactorization for minimizing network communication in graph computations. IEEE/ACM Trans. Networking **28**(6), 2768–2782 (2020)

A Class of Fast and Accurate Multi-layer Block Summation and Dot Product Algorithms

Kang He[1], Roberto Barrio[2], Lin Chen[1(✉)], Hao Jiang[3], Jie Liu[1], Tongxiang Gu[4], and Jin Qi[4]

[1] Science and Technology on Parallel and Distributed Processing Laboratory, National University of Defense Technology, Changsha 410073, China
{hekang2019,chen0,liujie}@nudt.edu.cn
[2] Department of Applied Mathematics, University of Zaragoza, E50009 Zaragoza, Spain
rbarrio@unizar.es
[3] College of Computer, National University of Defense Technology, Changsha 410073, China
haojiang@nudt.edu.cn
[4] Institute of Applied Physics and Computational Mathematics, Beijing 100000, China
{txgu,qi_jin}@iapcm.ac.cn

Abstract. Basic recursive summation and common dot product algorithm have a backward error bound that grows linearly with the vector dimension. Blanchard [1] proposed a class of fast and accurate summation and dot product algorithms respectively called FABsum and FABdot, which trades off the calculation accuracy and speed by the block size. Castaldo [2] proposed a multi-layer block summation and dot product algorithm called SuperBlocksum and SuperBlockdot that can increase the accuracy while adding almost no additional calculations. We combine the idea of [1] with the multi-layer block structure to propose SuperFABsum (for "super fast and accurate block summation") and SuperFABdot (for "super fast and accurate block dot product"). Our algorithms have two variants, one is SuperFAB(within), the other is SuperFAB(outside). Our algorithms further improve accuracy and speed compared with FAB and SuperBlock. We conducted accuracy and speed tests on the high-performance FT2000+ processor. Experimental results show that SuperFABdot(within) algorithm is more accurate than FABdot and SuperBlockdot. Compared with FABdot, SuperFABdot(outside) algorithm can achieve up to $1.2\times$ performance speedup while ensuring similar accuracy.

Keywords: Summation · Dot product · Accurate algorithm · Multi-layer block

This research is partly supported by the National Key Research and Development Program of China under Grant 2020YFA0709803, 173 Program under Grant 2020-JCJQ-ZD-029, the Science Challenge Project under Grant TZ2016002, and the Spanish Research project PGC2018-096026-B-I00 and the European Regional Development Fund and Diputación General de Aragón (E24-17R).

C. Cérin et al. (Eds.): NPC 2021, LNCS 13152, pp. 64–75, 2022.
https://doi.org/10.1007/978-3-030-93571-9_6

1 Introduction

Summation and dot product are the most basic tasks of floating-point arithmetic and are widely used in linear algebra systems. However, due to the accumulation of rounding errors, there will lead to loss of accuracy as increasing in computing scale. Dot product includes two types of calculations: summation and multiplication. The multiplication error in dot product will not accumulate, and hence the main algorithmic opportunity for error reduction is to choose a suitable summation algorithm [3–6]. Kahan [7] designed compensated summation to reduce errors. Ogita and Rump proposed error-free transformation technology in [8]. Researchers used this technology to design a series of high-precision numerical algorithms. However, these high-precision algorithms require more computational costs.

It is important to strike a balance between accuracy and speed. Blanchard and Higham proposed a class of fast and accurate summation and dot product algorithms in [1] respectively called FABsum and FABdot. Castaldo [2] proposed multi-layer block summation and dot product algorithms called SuperBlocksum and SuperBlockdot. SuperBlock algorithms permit researchers to make trade-offs between computational performance, memory usage, and rounding error. Unlike many such high-precision algorithms, SuperBlock algorithms require no additional floating-point operations while reducing rounding errors. However, the rounding error of both algorithms will increase rapidly when calculating ill-conditioned data.

We combine the idea of [1] with the multi-layer block structure in [2] to implement a class of fast and accurate multi-layer block summation and dot product algorithms called SuperFABsum and SuperFABdot. SuperFAB algorithms have two variants, one is to apply an accurate summation algorithm within the block and a fast summation algorithm outside the block, called SuperFAB(within). The other is to apply an accurate algorithm outside the block and a fast algorithm within the block called SuperFAB(outside). Experimental results show that the multi-layer block algorithm is faster than the unchunked algorithm when running in practice. SuperFABdot(within) algorithm is more accurate compared with FABdot and SuperBlockdot. Compared with the SuperBlockdot algorithm, SuperFABdot(outside) is also more accurate. Moreover, SuperFABdot(outside) needs to consume less computational time while maintaining the same accuracy as FABdot.

2 Previous Work

In this section, we begin by briefly reviewing existing summation algorithms and dot product algorithms. We use $fl(\cdot)$ to denote the evaluated result of an expression and introduce the standard model [9] of IEEE floating-point arithmetic. Note that this model does not account for the possibility of underflow (or overflow). In this paper, we assume that no overflow or underflow occurs. For floating-point numbers a, b, we have

$$fl(a \operatorname{op} b) = (a \operatorname{op} b)(1 + \delta_1), |\delta_1|, |\delta_2| \le u, \operatorname{op} \in \{+, -, \times, \div\}, \tag{1}$$

where u is the unit roundoff. This is the basic process of generating errors in floating-point computing. Higham [9] provides a particularly elegant way to perform rounding error analysis. We introduce two error analysis symbols θ_n and γ_n.

If $\delta_i \leq u$, $\rho_i = \pm 1$ and $nu < 1$, then

$$\prod_{i=1}^{n}(1+\delta_i)^{\rho_i} = 1 + \theta_n; \ |\theta_n| \leq \gamma_n = \frac{nu}{1-nu}. \tag{2}$$

2.1 Existing Summation Algorithms

Recursive Summation. Recursive summation is the most basic summation algorithm. The computed sum $\hat{s}$ satisfies [10] $\hat{s} = \sum_{i=1}^{n} x_i(1+\delta_i), |\delta_i| \leq \gamma_{n-1} = (n-1)u + O(u^2)$, where $\gamma_n = nu/(1-nu)$ for $nu < 1$. The backward error bound of Algorithm 1 grows linearly with n. When n is large enough, recursive summation cannot guarantee the accuracy of results.

Algorithm 1. Recursive summation [9]

This algorithm takes as input n summands x_i and it returns the sum $s = \sum_{i=1}^{n} x_i$.

1: $s = 0$
2: **for** $i = 1 : n$ **do**
3:     $s = s + x_i$
4: **end for**

Compensated Summation. For any pair (a, b) of floating-point numbers and their computed sum $\hat{s} = fl(a + b)$, there exists a floating-point number e such that $a + b = s + e$. Compensated summation evaluates the error term e at each step recursive summation, and compensates e back to original result. The backward error for compensated summation satisfies [8,10] $\hat{s} = \sum_{i=1}^{n} x_i(1+\delta_i)$, $|\delta_i| \leq 2u + O(u^2)$. This is an almost ideal result. Compared with the recursive summation, compensated summation requires 3 extra flops per loop iteration, which is typically more expensive than simply switching to higher precision.

Algorithm 2. Compensated summation [7]

This algorithm takes as input n summands x_i and computes their sum by compensated summation.

1: $s = 0, e = 0$
2: **for** $i = 1 : n$ **do**
3:     $temp = s$
4:     $y = x_i + e$
5:     $s = temp + y$
6:     $e = (temp - s) + y$
7: **end for**

FABsum. FABsum is a class of fast and accurate summation algorithms. The key idea of FABsum is to divide n summands into n/b blocks of size b, using the FastSum algorithm to compute the local sums s_i of each block, and using the AccurateSum algorithm to sum the local sum. When FastSum uses recursive summation and AccurateSum uses compensated summation, the backward error bound of FABsum with a fixed block size b satisfies [1] $\hat{s} = \sum_{i=1}^{n} x_i(1+\delta_i)$, $|\delta_i| \leq (b+1)u + O(u^2)$. Compared with recursive summation, the extra flops required for FABsum is a very small overhead for a typical block size selection, e.g., for $b = 300$, this represents an overhead of only 1% [1].

Algorithm 3. FABsum (Fast and accurate blocked summation algorithm) [1]

This algorithm takes as input n summands x_i a block size b, and two summation algorithms **FastSum** and **AccurateSum**. It returns the sum $s = \sum_{i=1}^{n} x_i$.

1: **for** $i = 1 : n/b$ **do**
2:     Compute $s_i = \sum_{j=(i-1)b+1}^{ib} x_j$ with **FastSum**
3: **end for**
4: Compute $s = \sum_{i=1}^{n/b} s_i$ with **AccurateSum**

2.2 Existing Dot Product Algorithms

In this section, we will give an overview of several dot product algorithms to compute the $s = \sum_{i=1}^{n} x_i y_i$.

Common Dot Product. Common Dot Product algorithm uses recursive summation algorithm (Algorithm 1) for summation. Common Dot Product has one additional error factor u due to the multiply $x_i y_i$, the computed result $\hat{s}$ satisfies [2] $\hat{s} = \sum_{i=1}^{n} x_i y_i(1+\delta_i)$, $|\delta_i| \leq nu + O(u^2)$.

SuperBlockdot Algorithm. Castaldo proposed SuperBlock dot product algorithm in [2]: Given t temporaries we can accumulate a sum in t levels, where t denotes the layers number of block; so we can find optimal blocking factors for each level to minimize the upper bound of the floating-point error. SuperBlockdot will provide reasonable error reduction on pretty much all problem sizes while requiring only one additional buffer. The computed result $\hat{s}$ of SuperBlockdot algorithm satisfies $\hat{s} = \sum_{i=1}^{n} x_i y_i(1 + \delta_i)$, $|\delta_i| \leq (2\sqrt{(n/b - 1)} + b)u + O(u^2)$.

Algorithm 4. Common Dot Product

This algorithm takes as input $x, y \in \mathbb{R}^n$. It returns the result $s = \sum_{i=1}^{n} x_i y_i$.

1: $s = 0$
2: **for** $i = 1 : n$ **do**
3:     $s = s + x_i y_i$
4: **end for**

FABdot Algorithm. FABdot algorithm is the dot product algorithm that uses FABsum algorithm for summation operations in [1]. When the size of b is selected appropriately, the balance of speed and accuracy can be implemented. If AccurateSum uses compensated summation, the backward error bound satisfies $\hat{s} = \sum_{i=1}^{n} x_i y_i(1 + \delta_i)$, $|\delta_i| \leq \varepsilon(n, b) = (b+2)u + O(u^2)$.

Algorithm 5. SuperBlockdot (SuperBlock Dot Product algorithm of t-layer block ($t = 3$)) [2]

This algorithm takes as input $x, y \in \mathbb{R}^n$ and a block size b, $t = 3$ denotes the structure of the three-layer block, and returns the result $s = \sum_{i=1}^{n} x_i y_i$. $cdot_j$ is local dot product in the block, $sdot_k$ is local dot product of the second layer of blocks, **layer_1** and **layer_2** are block parameters.

1: **layer_1** = sqrt(n/b)
2: **layer_2** = $(n/b)/$sqrt(n/b)
3: **for** $k = 1$: layer_1 **do**
4:     **for** $j = 1$: layer_2 **do**
5:         **for** $i = 1$: b **do**
6:             $cdot_j = cdot_j + x_i y_i$.
7:         **end for**
8:         Compute $sdot_k = \sum_{j=1}^{layer\_2} cdot_j$ with **Recursive summation**.
9:     **end for**
10:    Compute $s = \sum_{k=1}^{layer\_1} sdot_k$ with **Recursive summation**.
11: **end for**

Algorithm 6. FABdot (Fast and accurate blocked dot product algorithm) [1]

This algorithm takes as input $x, y \in \mathbb{R}^n$ and a block size b and returns the result $s = \sum_{i=1}^{n} x_i y_i$.

1: **for** $i = 1$: n/b **do**
2:     **for** $j = 1$: b **do**
3:         $s_i = s_i + x_j y_j$
4:     **end for**
5:     Compute $s = \sum_{i=1}^{n/b} s_i$ with **AccurateSum**.
6: **end for**

---

3   Multi-layer Block SuperFABsum and SuperFABdot Algorithms

Although SuperBlock algorithms and FAB algorithms can improve the accuracy, the error of their results will increase rapidly when calculating ill-conditioned data. So we combine the idea of [1] with the multi-layer block structure of SuperBlock to implement SuperFAB algorithms, which are more accurate. In this section, we will introduce the implementation and analysis of SuperFAB algorithms.

3.1   Implementation of SuperFABsum and SuperFABdot Algorithms

Compared with FAB algorithms, SuperFAB algorithms add adjustable multi-layer block structures. And compared with SuperBlock algorithms, SuperFAB algorithms implement the balance of accuracy and speed by using different accuracy inside and outside the block. The error of the dot product algorithm mainly comes from the summation operation. The dot product can be divided into two steps: multiplication and summation. The initial products $z_i = x_j y_j$ each incur one rounding error. So the backward error bound of the dot product algorithm is only one unit roundoff more than the summation algorithm used. In this section, we mainly introduce the implementation and analysis of the SuperFABdot algorithm.

SuperFABdot algorithms have two variants, one is to apply an accurate summation algorithm (such as compensated summation, or recursive summation in higher precision) within the block and a fast summation algorithm (such as recursive summation) outside the block, called SuperFAB(within) (Algorithm 7). The other is to apply an accurate algorithm outside the block and a fast algorithm within the block, called Super-FAB(outside) (Algorithm 8). SuperFABdot algorithm can also weigh the accuracy and speed of the calculation with a suitable choice of block size b. When layer_2 $= 1$, SuperFABdot(outside) algorithm is the same as FABdot algorithm. SuperFABdot algorithm is the same as SuperBlockdot algorithm when both inside and outside the block are FastSum.

Algorithm 7. SuperFABdot(within) algorithm of t-layer block ($t = 3$)

This algorithm takes as input $x, y \in \mathbb{R}^n$ and a block size b, $t = 3$ denotes the structure of the three-layer block, and returns the result $s = \sum_{i=1}^{n} x_i y_i$. $cdot_j$ is local dot product in the block, $sdot_k$ is local dot product of the second layer of blocks, **layer_1** and **layer_2** are block parameters.

1: **layer_1** = sqrt(n/b)
2: **layer_2** = $(n/b)/$ sqrt(n/b)
3: **for** $k = 1$: layer_1 **do**
4:    **for** $j = 1$: layer_2 **do**
5:       **for** $i = 1$: b **do**
6:          Compute $cdot_j = \sum_{i=1}^{b} x_i y_i$ with **AccurateSum**.
7:       **end for**
8:       Compute $sdot_k = \sum_{j=1}^{layer\_2} cdot_j$ with **FastSum**.
9:    **end for**
10:   Compute $s = \sum_{k=1}^{layer\_1} sdot_k$ with **FastSum**.
11: **end for**

3.2 Rounding Error Analysis

Now, we analyze the error bound of the SuperFABdot algorithm. For any dot product of the form $s = \sum_{i=1}^{n} x_i y_i$, we assume that the computed $\hat{in}$ of computing in the block satisfies $\hat{in} = \sum_{i=1}^{n} x_i y_i (1 + \delta_i^b)$, $\left|\delta_i^b\right| \leq \varepsilon_b(n)$, and the computed $\hat{out}$ of computing outside the block satisfies $\hat{out} = \sum_{j=1}^{n/b} s_j (1 + \delta_j^a)$, $\left|\delta_i^a\right| \leq \varepsilon_a(n)$, where $\varepsilon_a(n)$ is the error within block and $\varepsilon_b(n)$ is the error outside block. With these assumptions, firstly we analyze the error bound of SuperFAB(within) algorithm.

- If AccurateSum uses recursive summation in extended precision, with a final rounding back to the working precision, then we have $\varepsilon_b(n) = 2u + O(u^2)$, $\varepsilon_a(n/b) = 2(\sqrt{n/b} - 1) + O(u^2)$. So the forward error bound of SuperFABdot(within) algorithm (Algorithm 7) satisfies

$$|s - \hat{s}| \leq \gamma_{(2\sqrt{n/b})} |x|^T |y|. \tag{3}$$

Algorithm 8. SuperFABdot(outside) algorithm of t-layer block ($t = 3$)

This algorithm takes as input $x, y \in \mathbb{R}^n$ and a block size b, $t = 3$ denotes the structure of the three-layer block, and returns the result $s = \sum_{i=1}^{n} x_i y_i$. $cdot_j$ is local dot product in the block, $sdot_k$ is local dot product of the second layer of blocks, **layer_1** and **layer_2** are block parameters.

```
1:  layer_1 = sqrt(n/b)
2:  layer_2 = (n/b)/ sqrt(n/b)
3:  for k = 1 : layer_1 do
4:      for j = 1 : layer_2 do
5:          for i = 1 : b do
6:              Compute cdot_j = ∑_{i=1}^{b} x_i y_i with FastSum.
7:          end for
8:          Compute sdot_k = ∑_{j=1}^{layer_2} cdot_j with AccurateSum.
9:      end for
10:     Compute s = ∑_{k=1}^{layer_1} sdot_k with AccurateSum.
11: end for
```

At this point, the forward error bound of SuperFABsum(within) algorithm satisfies

$$|s - \hat{s}| \leq \gamma_{(2\sqrt{n/b}-1)} |x|^T |y| . \tag{4}$$

– If AccurateSum uses compensated summation, we have $\varepsilon_b(b) = 3u + O(u^2)$ and the forward error bound of SuperFABdot(within) algorithm satisfies

$$|s - \hat{s}| \leq \gamma_{(2\sqrt{n/b}+1)} |x|^T |y| . \tag{5}$$

The forward error bound of SuperFABsum(within) algorithm satisfies

$$|s - \hat{s}| \leq \gamma_{(2\sqrt{n/b})} |x|^T |y| . \tag{6}$$

According to the above analysis, we can get that the backward error bound of Super-FABdot(within) algorithm is significantly lower than the SuperBlock Dot Product algorithm in [2], and as long as we choose the block size $b > \sqrt[3]{4n}$, its backward error bound is also lower than FABdot algorithm in [1]. We can choose the appropriate block size b to weigh the accuracy and speed of calculation, to get better accuracy and speed. Next, we analyze the error bound of the SuperFABdot(outside) algorithm.

– If AccurateSum uses recursive summation in extended precision in SuperFAB-dot(outside) algorithm, with a final rounding back to the working precision, then we have $\varepsilon_b(n) = bu + O(u^2)$, $\varepsilon_a(n/b) = 2u + O(u^2)$. So the forward error bound of SuperFABdot(outside) algorithm (Algorithm 8) satisfies

$$|s - \hat{s}| \leq \gamma_{(b+2)} |x|^T |y| . \tag{7}$$

The forward error bound of SuperFABsum(outside) algorithm satisfies

$$|s - \hat{s}| \leq \gamma_{(b+1)} |x|^T |y| . \tag{8}$$

– If AccurateSum uses compensated summation, we have $\varepsilon_a(n/b) = 4u + O(u^2)$ and
the forward error bound of SuperFABdot(outside) algorithm satisfies

$$|s - \hat{s}| \leq \gamma_{(b+4)} |x|^T |y|. \tag{9}$$

The forward error bound of SuperFABsum(outside) algorithm satisfies

$$|s - \hat{s}| \leq \gamma_{(b+3)} |x|^T |y|. \tag{10}$$

Compared with the SuperBlock Dot Product algorithm, the error bound of the
SuperFABdot(outside) algorithm has been significantly reduced. Moreover, although
the error bound of the SuperFABdot(outside) algorithm has been not improved compared to the FABdot algorithm, its multi-layered block structure can reasonably use the
multi-level cache of modern processors and improve the cache hit rate. What needs to
be emphasized is modern architecture machines have the expansion of standard work-
stations to 64-bit memories and multicore processors, larger computations are possible
on even simple desktop machines than ever before. SuperFAB algorithms can make full
use of the various levels of cache on the computer and is more suitable for matrix multiplication computation on parallel machines, which will be a better computation speed.
Taking the SuperFABdot(within) algorithm as an example, we can perform SIMD
instruction-level parallel design within a block executing precise summation operations,
perform MPI data-level parallel design in the first-level block structure, and perform
OpenMP thread-level parallel design in the second block structure, to obtain a multi-
level parallel accurate dot product algorithm. The structure diagram of the multi-level
parallel algorithm is shown in Fig. 1, where N denotes the input vector dimension, N_p
denotes the sub-vector block dimension on each process, and N_t denotes the sub-vector
block dimension on each thread.

4   Numerical Experiments

We conducted accuracy and performance tests on the high-performance FT2000+ processor. The core processor parameter information is shown in Table 1. First, we experiment with IEEE single precision and double precision. Ogita provided Gendot algorithm to generate vectors x, y with anticipated condition number c in [8]. We use
ill-conditioned vectors generated by Gendot algorithm as input data and compute
$s = \sum_{i=1}^{n} x_i y_i$. For each test, we randomly generate 1,000 different vector pairs. We
choose the results computed by the high-precision floating-point computation library
MPFR [11] as an exact result and calculate the average relative error(i.e., the relative
error computation method is that for each input X and Y, we compute $\frac{|X \cdot Y - \overline{X \cdot Y}|}{|X \cdot Y|}$).
The names corresponding to the algorithms tested are shown in Table 2.

These parameter values are chosen due to practical reasons: $b = 60$ is a typical
midrange blocking factor (when b is selected for performance), and $t = 3$ requires
only one additional workspace. Figure 2a and Fig. 2b show the accuracy-test results
under single precision and double precision respectively. In Fig. 2a, the results of AccurateSum using compensated summation are the same as AccurateSum using recursive
summation in double precision, so we give only one result. In Fig. 2b, since the existing

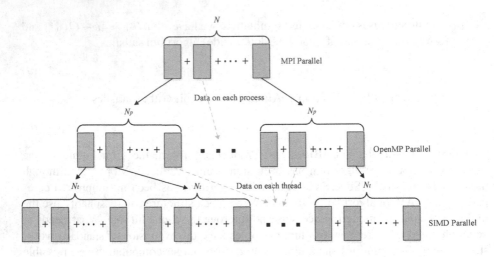

Fig. 1. Multi-level parallel algorithm structure diagram

Table 1. Core parameters Of FT2000+ processor

Base clock	2.2 GHz
Core	64 (1 CPU)
Cache	L1I: 48 KB L1D: 32 KB L2: 32 M
Main memory	8 * 8G
Memory	DDR4 2400 MHz
Compiler version	GCC-8.3.0
Compiler optimization options	-O3

hardware does not support scaling to quadruple precision, we do not show the results of AccurateSum using recursive summation in extended precision. From Fig. 2, we can see that SuperFABdot(within) has a lower relative error and is more accurate than FABdot and SuperBlockdot.

The next experiment uses IEEE half precision (fp16, $u = 2^{-11}$), bfloat16 ($u = 2^{-8}$) in MATLAB. We use the rounding function chop.m [12] in MATALAB to simulate low precisions to test the algorithm which AccurateSum uses recursive summation in extended precision $u_e = 2^{-24}$ (that is single precision). We choose the calculation result with double precision as the exact result. As can be seen from Fig. 3 and Fig. 4, SuperFABdot(within) algorithm is more accurate than other algorithms. Furthermore, SuperFABdot(outside) has similar accuracy as FABdot. Through the above experiments, we verified the previous analysis of the error bound.

Table 2. Names corresponding to the algorithms tested

Name	Algorithm
Commondot	Algorithm 4
SuperBlockdot	Algorithm 5 in [2]
FABdot	Algorithm 6 (AccurateSum uses compensated summation) in [1]
FABdotextended	Algorithm 6 (AccurateSum uses recursive summation in extended precision) in [1]
SuperFABdot(within)	Algorithm 7 (AccurateSum uses compensated summation)
SuperFABdotextended(within)	Algorithm 7 (AccurateSum uses recursive summation in extended precision)
SuperFABdot(outside)	Algorithm 8 (AccurateSum uses compensated summation)
SuperFABdotextended(outside)	Algorithm 8 (AccurateSum uses recursive summation in extended precision)

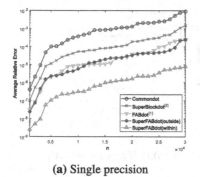

(a) Single precision

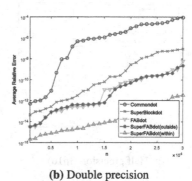

(b) Double precision

Fig. 2. Average relative error for the dot product $\sum_{i=1}^{n} x_i y_i$ computed in single and double precision, and block size $b = 60$, 1,000 Trials per Point per Algorithm.

We test the performance of different algorithms when working precision is single precision. Since the running time of algorithms will be affected by many non-controllable factors during the running process, for each pair of test vectors, we test 10,000 times to calculate the total running time $time_{test}$. We use the running time of Commondot as the baseline for testing, calculate the time rate ($rate = \frac{time_{test}}{time_{common}}$). Then we evaluate performance by comparing the time rates ($rate$) of different algorithms.

As can be seen from Fig. 5a, when AccurateSum uses compensated summation, the time rate of SuperFABdot(with) algorithm to Commondot algorithm is about 4.5. This is because SuperFABdot(with) algorithm adds high-precision calculation operations, which requires more computational cost. The time rate of the SuperFABdot(outside) algorithm compared to Commondot is 1.5, while the FABdot algorithm is 1.8. So our SuperFABdot(outside) algorithm can achieve up to $1.2\times$ ($speedup = \frac{rate_{SuperFABdot(outside)}}{rate_{FABdot}}$) performance speedup as while ensuring similar accuracy compared to FABdot algorithm developed by Blanchard and Higham. When AccurateSum is the recursive

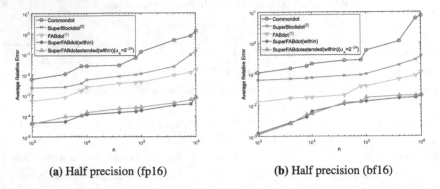

(a) Half precision (fp16)   **(b)** Half precision (bf16)

Fig. 3. Average relative error of using SuperFABdot(within) to computed $s = \sum_{i=1}^{n} x_i y_i$. For a vector x with random uniform $[0, 1]$ entries block size $b = 128$.

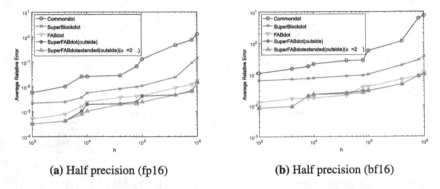

(a) Half precision (fp16)   **(b)** Half precision (bf16)

Fig. 4. Average relative error of using SuperFABdot(outside) to computed $s = \sum_{i=1}^{n} x_i y_i$. for a vector x with random uniform $[0, 1]$ entries block size $b = 128$.

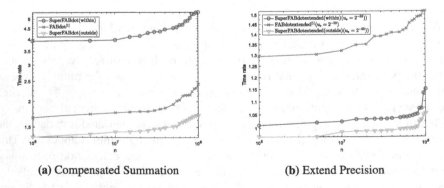

(a) Compensated Summation   **(b)** Extend Precision

Fig. 5. The calculation time rate of different dot product algorithms under different data scales in single precision, block sizes $b = 128$

summation of extending to double precision, SuperFABdotextended(outside) and SuperFABdotextended(with) algorithm have a calculation time close to that of Commondot. We speculate that the reason for acceleration is that the multi-layered block structure will guide the compiler to generate vectorized code and optimize it.

5 Conclusion

With the increasing floating-point computing scale, the error of common dot product and recursive summation algorithm will increase rapidly. Based on this situation, we combine the idea of the FAB algorithm with the multi-layer block structure to propose SuperFABsum and SuperFABdot algorithm. When AccurateSum uses compensated summation, the backward error of SuperFABdot(within) algorithm is $(2\sqrt{n/b}+1)u + O(u^2)$ and the backward error of SuperFABdot(outside) algorithm is $(b+4)u + O(u^2)$. The experimental results show that compared with the FAB-dot [1] and SuperBlockdot [2] dot product algorithms, SuperFABdot(within) algorithm is more accurate. Compared with FABdot, our SuperFABdot(outside) algorithm can achieve up to $1.2\times$ performance speedup while ensuring similar accuracy.

References

1. Blanchard, P., Higham, N.J., Mary, T.: A class of fast and accurate summation algorithms. SIAM J. Sci. Comput. **42**(3), A1541–A1557 (2020)
2. Castaldo, A.M., Whaley, C.R., Chronopoulos, A.T.: Reducing floating point error in dot product using the superblock family of algorithms. SIAM J. Sci. Comput. **31**(2), 1156–1174 (2009)
3. Gregory, J.: A comparison of floating point summation methods. Commun. ACM **15**(9), 838 (1972)
4. Linnainmaa, S.: Analysis of some known methods of improving the accuracy of floating-point sums. BIT Numer. Math. **14**(2), 167–202 (1974)
5. Robertazzi, T.G., Schwartz, S.C.: Best "ordering" for floating-point addition. ACM Trans. Math. Softw. (TOMS) **14**(1), 101–110 (1988)
6. Higham, N.J.: The accuracy of floating point summation. SIAM J. Sci. Comput. **14**(4), 783–799 (1993)
7. Kahan, W.: Pracniques: further remarks on reducing truncation errors. Commun. ACM **8**(1), 40 (1965)
8. Ogita, T., Rump, S.M., Oishi, S.: Accurate sum and dot product. SIAM J. Sci. Comput. **26**(6), 1955–1988 (2005)
9. Higham, N.J.: Accuracy and Stability of Numerical Algorithms, 2nd edn. SIAM, PA, USA (2002)
10. Goldberg, D.: What every computer scientist should know about floating-point arithmetic. SIAM J. Sci. Comput. **23**(1), 5–48 (1991)
11. Zimmermann, P.: Reliable computing with GNU MPFR. In: International Congress on Mathematical Software, pp. 42–45 (2010)
12. Higham, N.J., Pranesh, S.: Simulating low precision floating-point arithmetic. SIAM J. Sci. Comput. **41**(5), C585–C602 (2019)

A KNN Query Method for Autonomous Driving Sensor Data

Tang Jie[1], Zhang Jiehui[1], Zeng Zhixin[1(✉)], and Liu Shaoshan[2]

[1] South China University of Technology, Guangzhou, China
cstangjie@scut.edu.cn, cszzx@mail.scut.edu.cn
[2] PerceptIn, Santa Clara, U.S.A

Abstract. Autonomous driving cars need to perceive and extract environmental information through sensors around the body during the driving process. Sensor data streams are rich in semantic information, and semantic annotation can form semantic text with geographic location information. This information is important for scenario applications such as effective updating of high-precision maps and more accurate perception of the surrounding environment. Therefore, it is necessary to store and organize them effectively. At the same time, users' demand for such information retrieval is increasing, and it is more and more important to efficiently retrieve useful information for users from the huge amount of information. In this paper, We propose an index, SIR-tree, to efficiently organize the spatial, textual and social information of objects. SIR-tree is an R-tree containing both social relevance and textual relevance, and each parent node in the tree contains the spatial, textual and social information of all children nodes, and has good scalability. Based on the SIR-tree index we propose the priority traversal algorithm *BF*, which uses a priority queue to store all candidate objects and traverses the nodes to filter them according to their ranking function values. To optimize the query algorithm, we propose two pruning strategies, distance-based *BF* algorithm and X-HOP_BF algorithm, to improve the query efficiency. Experimental results show that the *BF* algorithm takes at least 70 s to return 20 objects in the Gowalla dataset, while the X-HOP_BF algorithm only takes about 20 s.

1 Introduction

With the increasing popularity of location-based social networking services, social networks have become a platform and tool for people to communicate in their daily lives. Users' friendships on social networks largely reflect their relationships in real life. However, existing spatial keyword retrieval techniques focus only on the distance constraint between the user and the object, which becomes limited in the spatial keyword retrieval scenario containing social network information. Imagine the following scenario: suppose user A takes his location as the query center point, the query keyword input is coffee, and the total number of query k input is 2. If we only consider the spatial distance and keyword, then the two closest coffee shops a and b will be returned, but consider the following situation: user A has clocked in and checked in many times at coffee shops p3, p4, and coffee shop p3 also has a number of visits by user A's friends user B and user C. While coffee shops p1 and p2 are close but not visited by user A's social friends, this suggests that the user may prefer p3 because not only does he go there often, but his friends also go there often, even if they are a little further away. Therefore, the query result will return {p3} because it takes into account both spatial and textual constraints as well as social relevance and will better satisfy the user's query requirements.

To solve this problem, we first design a rank function to measure the relevance between the query set and the candidate set. Next, to improve query efficiency, we develop a novel hybrid index structure, SIR-tree, which integrates social, spatial and textual properties.

Specifically, we have made the following contributions.

1. We introduce the concept of social relevance. The property of a fan set is added to each spatial-text object, and each fan in the fan set has a positive attitude towards the object. The social relevance of an object depends on the relationship between the querying user and all the fans of the object in the social network.
2. We propose a hybrid index structure based on IR-tree, called "Socially aware IR-tree (SIR-tree)", which integrates location, text and social information, and we present construction and update algorithms for this index.
3. We propose a *BF* traversal algorithm based on the total object score rank(p), and two improved algorithms based on *BF* to further prune the objects, called *DIST_BF* algorithm based on distance pruning strategy and X-HOP_BF algorithm based on social distance pruning strategy, respectively
4. We conducted diverse experiments on two datasets, Brightkite and Gowalla, to evaluate the correctness of our proposed optimization algorithms.

The remainder of the paper is organized as follows: first we discuss related work in Sect. 2, then in Sect. 3 we propose a novel index structure SIR-tree for spatial keyword queries with social relevance. In Sect. 4 we propose an efficient pruning algorithm to obtain top k results. Section 5 conducts rich experiments. And we conclude with conclusions.

2 Related Work

Since spatial-text queries may have various conditional constraints (e.g., time, personal preferences, road network restrictions, etc.), this poses a great challenge for organizing the indexing structure of spatial objects. A large number of researchers have investigated the problem and proposed solutions. To support online spatial keyword queries, developers usually deploy data and query processing services on cloud platforms, but this solution may cause privacy leakage problems. The literature [1] presents and normalizes the problem based on privacy-preserving Boolean spatial keyword queries. A privacy-preserving spatial text filter encoding structure and an encrypted R-tree index are designed. A novel framework to solve the top-k location-aware similarity search problem with fuzzy text matching is proposed in the literature [2]. Firstly, a hierarchical index HGR-Tree is proposed to capture the signatures of spatial and textual relevance. Based on such an index structure, Yang et al. designed a best-first search algorithm that can preferentially access HGR-Tree nodes with more similar objects, while pruning objects from HGR-Tree nodes with dissimilar objects. Based on this, an incremental search strategy is further designed to reduce the overhead caused by supporting fuzzy text matching. As a major type of continuous spatial query, mobile spatial keyword queries have been extensively studied. Most of the existing studies have focused on retrieving individual objects, each of which is close to the query object and related to the query keywords. However, a single object may not satisfy all the needs of the user, for example, a user who is driving may want to pick up money, wash the car, and buy some medicine, which can only be satisfied by multiple objects. A new query type called Moving Collective Spatial Keyword Query (MCSKQ) is proposed in the literature [3]. The literature [4] continues the study of mobile objects by proposing dynamic spatial keyword object types where location and keywords change over time. The problem of tracking the first k dynamic spatial keyword objects consecutively for a given query is also investigated, and a grid index-based indexing structure is proposed. Carmel et al. investigated personalized social search based on the user's social relations. Search results are re-ranked according to their relations with individuals in the user's social network. SonetRank personalized the Web search results based on the aggregate relevance feedback of the users in similar groups, based on the observation that users in the same group may have similar relevance judgments for queries related to these groups. Xu et al. and proposed a ranking method based on tag-relevance and match between users' profiles anddocument tags.

3   SIR-Tree

3.1   Problem Definition

Spatial-Text Objects: Given a dataset of spatial-text objects $\mathcal{D}$, the $\mathcal{D}$ the spatial-text objects in p $(p \in \mathcal{D})$ is a triplet $\langle \delta, \varphi, \mathcal{M} \rangle$. $p.\delta, p.\varphi, p.\mathcal{M}$ The objects are, respectively p of location points, text descriptions, and the set of fans. A fan u $(u \in p.\mathcal{M})$ is a user who has expressed a positive attitude towards the object p user who has expressed a positive attitude towards the object.

Socially-Aware Top-k Space-Based Keyword Query: Given a socially-aware top-k space-based keyword query $(SKSK)$ $q = \langle \delta, \varphi, k, S \rangle$, the $q.\delta, q.\varphi, q.k, q.S$ are the query q the query location point, the set of query keywords, the number of objects to be returned, and the social network of the user making the query. A social network S is abstracted as an undirected graph, where each node represents a user and each edge represents a relationship. The social distance between users u_1, u_2 The social distance between $dist_s(u_1, u_2)$ is defined as the shortest path between the u_1, u_2.

Ranking Score: A $SKSK$ query returns a list of the k objects with the largest ranking scores, and the objects in this list are sorted in descending order of ranking score, which is defined as follows.

$$rank_q(p) = \varepsilon * dp_q(p) + \beta * tr_q(p) + (1 - \varepsilon - \beta) * sd_q(p) \tag{1}$$

Balance parameters $\varepsilon, \beta \in [0, 1]$ and $\varepsilon + \beta \in [0, 1]$. $dp_q(p)$ indicates that the query user q and the object p the spatial proximity between $dp_q(p) = 1 - \frac{dist(q,p)}{D_{max}}$. The function $tr_q(p)$ denotes the textual correlation between the query q and object p, and this paper uses the TF-IDF model. The function $sd_q(p)$ represents the social relevance between the query q and object p.

Social Relevance: The use of social relevance facilitates querying objects in social networks that intersect more with the user's social network, and the social relevance function $sd_q(p)$ is defined as.

$$sd_q(p) = \sum_{u \in p.M} \alpha^{dist_s(u_q, u)} \tag{2}$$

Parameters $\alpha \in [0, 1]$, the relevance calculation model applied by the algorithm in this paper is not limited to the model proposed above, other similarity calculation methods are still applicable. We apply the above concepts to a practical example. Figure 1 illustrates the set of object $\mathcal{D} = \{p_1, p_2, p_3, p_4\}$ of location points. The query points in the figure q of the set of query keywords $q.\varphi = \{a, b\}, q.k = 2$. The top right of the object $\langle . \rangle$ in the figure is the text relevance calculated using $tr_q(p)$ calculated text relevance. Let the query user be u_1 and their social network $q.S$ as shown in Fig. 2. The contents in Table 1 are the query location $q.\delta$ the Euclidean distance to the object and the set of followers of the object.

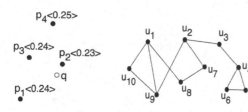

p_4<0.25>

p_3<0.24> p_2<0.23>

○q

p_1<0.24>

Fig. 1. Spatial distribution   **Fig. 2.** Users' social network

Table 1. Query point-to-object distance and the set of fans of the object

	$dist(q, p_i)$	$p_i.\mathcal{M}$
p_1	0.14	u_2, u_7
p_2	0.11	u_1, u_3, u_9
p_3	0.13	u_5, u_6
p_4	0.14	u_7, u_{10}

3.2   Indexing Model

To improve the query efficiency, a hybrid spatial-textual index structure, the socially aware IR-tree (SIR-tree), is proposed. This index structure can compute Euclidean distance, textual relevance, and social relevance simultaneously. Because it is an R-tree-based structure, the index can use the boundary scores of non-leaf nodes to prune the search space. The SIR-tree is extended from the IR-tree [6], where each node in the IR-tree has a collection of fans for each entry, and these collections are the concatenation of the sets of object fans contained in the subtree of that entry. Suppose there is a node in the SIR-tree n in which there are three entries n_1, n_2, n_3 and their fan sets are $n_1.\mathcal{M}$ and $n_2.\mathcal{M}$ and $n_3.\mathcal{M}$. Therefore the node n The parent node of n_p of the set of fans is $n_p.\mathcal{M} = \bigcup_{i=1,2,3} n_i.\mathcal{M}$.

Leaf Nodes in the SIR-Tree: In the SIR-tree, each leaf node contains many entries of the form $\langle ptr, MBR, \varphi, \mathcal{M} \rangle$ of entries. ptr Pointing to an object in the dataset $\mathcal{D}$ objects in the p, MBR is the object p of the smallest bounding rectangle. The minimum bounding rectangle is used to approximate more complex shapes, where the sides of the rectangle are parallel to the x-axis and y-axis and are as close as possible to the shape. Figure 3 shows the MBR for a complex shape. φ is the identifier of the document and $\mathcal{M}$ is the object p of the fan collection. Each leaf node also contains a pointer to an inverted document, which is used to store the text of the objects in the node.

Non-Leaf Nodes in the SIR-Tree: Each non-leaf node N contains many entries of the form $\langle ptr, MBR, \varphi, \mathcal{M} \rangle$ entries, where ptr is a pointer to a child node, and MBR is the smallest bounding rectangle of all child nodes of N, and φ is a reference to all pseudo documents in the child node entries, which can be used to derive the upper bound of textual relevance. Each non-leaf node also contains an inverted document pointer to the pseudo document of the entry stored in that node.

Figure 3 shows a SIR-tree with 9 spatial-text objects, and Fig. 4 shows the contents of the inverted document associated with the nodes. The nodes R_5 entry of the R_2 in the node has a weight of 7 for word c, which is the R_2 in the maximum weight of the words in the three documents. Figures 2 and 3 illustrates how the set of fans of an entry is calculated and stored in the SIR-tree: the set of fans of the object p_1, p_2 of the fan set are $\{u_1, u_3, u_5\}$ and $\{u_3, u_5\}$. Their parent node R_1 is the set of fans of p_1, p_2 the concatenation of the set of fans, i.e. $\{u_1, u_3, u_5\}$.

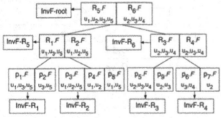

$InvF$-root		$InvF$-R_5		$InvF$-R_6	
a: $(R_5, 7), (R_6, 4)$		a: $(R_1, 5), (R_2, 7)$		a: $(R_3, 4), (R_4, 1)$	
b: $(R_5, 5), (R_6, 4)$		b: $(R_1, 5), (R_2, 3)$		b: $(R_4, 4)$	
c: $(R_5, 7), (R_6, 4)$		c: $(R_1, 5), (R_2, 7)$		c: $(R_3, 4), (R_4, 4)$	
d: $(R_5, 1), (R_6, 1)$		d: $(R_2, 1)$		d: $(R_4, 1)$	

$InvF$-R_1	$InvF$-R_2	$InvF$-R_3	$InvF$-R_4
a: $(p_1, 5)$	a: $(p_3, 7)$	a: $(p_5, 4), (p_9, 3)$	a: $(p_7, 1)$
b: $(p_2, 5)$	b: $(p_8, 3)$		b: $(p_6, 4), (p_7, 1)$
c: $(p_1, 5), (p_2, 5)$	c: $(p_4, 7), (p_8, 3)$	c: $(p_5, 4), (p_9, 3)$	c: $(p_6, 3), (p_7, 4)$
	d: $(p_3, 1), (p_4, 1)$		d: $(p_7, 1)$

Fig. 3. Structure of SIR-tree   **Fig. 4.** Inverted document with node pointing

Since the SIR-tree is extended on the structure of the IR-tree, the textual relevance of non-leaf node entries is an upper bound on the textual relevance of its child nodes, and Theorem 1 formally states this property.

Theorem 1: Given a query q and a non-leaf entry e that e the child node of contains m entries $CE = \{ce_i, 1 \leq i \leq m\}$, then $\forall ce_i \in CE\{tr_q(ce_i) \leq tr_q(e)\}$

For example, given a query q that has $tr_q(R_5) \geq tr_q(R_1) \geq tr_q(p_1)$.

Theorem 2: Given a query q and a non-leaf entry e that e the child node of contains m entries $CE = \{ce_i, 1 \leq i \leq m\}$, then $\forall ce_i \in CE\{sd_q(ce_i) \leq sd_q(e)\}$

Proof: Because ce_i In the e the child nodes of ce_i the set of fans of e the subset of $ce_i.\mathcal{M} \subseteq e.\mathcal{M}$, according to the social relevance $sd_q(.)$ the definition of has $sd_q(ce_i) = \sum_{u \in ce_i.\mathcal{M}} \alpha^{dist_s(u_q, u)} \leq \sum_{u \in e.\mathcal{M}} \alpha^{dist_s(u_q, u)} = sd_q(e)$, proving Bi.

Theorem 3: Given a query q and a non-leaf entry e that e the child node of contains m entries $CE = \{ce_i, 1 \leq i \leq m\}$, then $\forall ce_i \in CE\{rank_q(ce_i) \leq rank_q(e)\}$

Proof: By definition q to e the Euclidean distance is less than or equal to q to ce_i the distance to the Euclidean distance, and from Theorems 1, 2 we know that $tr_q(ce_i) \leq tr_q(e)$ that $sd_q(ce_i) \leq sd_q(e)$, and $rank_q(ce_i) = \alpha * dp_q(ce_i) + \beta * tr_q(ce_i) + (1 - \alpha - \beta) * sd_q(ce_i) \leq \alpha * dp_q(e) + \beta * tr_q(e) + (1 - \alpha - \beta) * sd_q(e) = rank_q(e)$, the proof is over.

3.3 Building and Updating of the SIR-Tree

Next, we propose the construction algorithm of SIR-tree. The construction process of SIR-tree is also the process of node insertion.

The main idea of the algorithm is as follows: (line 1) First call the ChooseLeaf() function to select a leaf node that is most suitable for insertion into the object N. (lines 2–3) Update the MBR, inverted document, and fan set of N. (Lines 5–6) If the node needs to be split, call the split function to split the node. (Lines 7–11) If N is the root node, a new root node P needs to be created as the root node and N1, N2 as the children nodes. (Lines 12–16) If N is not the root node, then update the MBR, inverted document, and fan set of the ancestor node up to the root node. If the node does not need to split, the MBR, inverted document, and fan set of the ancestor node are updated directly. The pseudo code of the insertion algorithm is shown below.

Algorithm 1: SIR-tree of Node Insertion

```
1:  function INSERT(loc, document, M)
2:      N←ChooseLeaf(loc, RootNode)
3:      add loc to node N, add document to the inverted file of N
4:      update N.M to the union of M and N.M
5:      if N needs to be split then
6:          {N1, N2} ← N.split
7:          if N is root node then
8:              Initialize a new node P
9:              Add N1 and N2 to node P
10:             update the invert file of N1, N2 and P
11:             Set P to the root node
12:         else
13:             Ascend from N to the root
14:             adjusting covering rectangles
15:             updating the inverted file and the set of fans
16:         end if
17:     else if N is not root then
18:         Update the covering rectangles
19:         inverted files and fans sets of the ancestor nodes of N
20:     end if
21: end function
```

4 KNN Query Algorithm

4.1 Traversal Algorithm *BF* Based on Total Object Score

To process *SKSK* queries using the SIR-tree, we use the best-first traversal (*BF*) algorithm to retrieve the first k objects. Using a priority queue to keep the objects that have not yet been accessed, with the key of each object being the function $rank_q(e')$ value, the larger the value, the higher the position in the queue. the larger the value, the higher the position in the queue. The pseudo code of this algorithm is shown in Algorithm 2. (Lines 2–3) First initialize a priority *Queue* that holds all candidate objects (nodes or spatial objects) and puts the root node of the SIR-tree into the priority queue. (Lines 4–5) When *Queue* we enter the loop and return the first object in the priority queue when there are still candidates in the queue.*e*. (Lines 6–9) If *e* is a space object, then we add *e* to the result list, and the algorithm ends if the number of spatial objects in the result list reaches k. (Lines 10–15) *e* is a node, traverse *e* each child node in *e'*, using breadth-first search to determine the social relevance of the visited nodes $sd_q(e')$. Calculate the *e'* the ranking function of $rank_q(e')$ that will $rank_q(e')$ as *e'* the key value is inserted into the priority *Queue* in the queue.

Algorithm 2: BEST-FIRST Traversal Algorithm

```
1:  function SIRTQ(Query q, Integer k, SIR-tree index)
2:      Queue ← NewPriorityQueue();
3:      Queue.Enqueue(index .RootNode, 0 );
4:      while not Queue.IsEmpty() do
5:          Entry e ← Queue.Dequeue();
6:          if e refers to an object then
7:              Add e to the top-k result;
8:              if k objects have been found then return the top-k result;
9:          end if
10:         else
11:             for each entry e' in node e do
12:                 Compute its social relevance sd_q (e') using BFS;
13:                 Queue.Enqueue (e', rank_q (e'))
14:             end for
15:         end if
16:     end while
17: end function
```

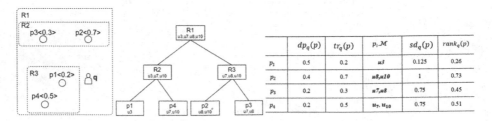

	$dp_q(p)$	$tr_q(p)$	$p_i \cdot \mathcal{M}$	$sd_q(p)$	$rank_q(p)$
p_1	0.5	0.2	$u3$	0.125	0.26
p_2	0.4	0.7	$u8,u10$	1	0.73
p_3	0.2	0.3	$u7,u8$	0.75	0.45
p_4	0.2	0.5	u_7, u_{10}	0.75	0.51

(a) A SkSk query   (b) The SIR-tree of example   (c) The score of object P

Fig. 5. An example of the SkSK queries

Example: Assume that user u_1 conducts a top 1 query of SKSK at point q, assume $\alpha = 0.3$, $\beta = 0.3$, $(1 - \alpha - \beta) = 0.4$, and let the parameter α in $sd_q(p)$ be 0.5. Firstly, we take the root node R_1 of SIR-Tree from the priority queue, and calculate their social relevance score and total score for each child node of R_1, namely R_2 and R_3. For node R_2, its fan set is the union of sub-nodes. According to social network Fig. 2, we define: For each edge added, social distance increases by 1, so $sd_q(R2) = 0.5^3 + 0.5^2 + 0.5^1$, $sd_q(e)$ and $tr_q(e)$ are the maximum of all sub-nodes, so the total score of R_2 is $rank_q(R_2) = 0.3 * 0.5 + 0.3 * 0.5 + 0.4 * 0.75 = 0.6$. In the same way, we can calculate $rank_q(R_3) = 0.73$. Because priority queues are sorted according to $rank_q$, in the second round, we first select node R_3 with a higher total score, calculate sd_q and $rank_q$ of objects $p2$ and $p3$ in the same way, and then put them into the priority queue. In the third round, $p2$ with the highest total score is first selected, because P2 is an object, and it is added to the result set. At this point, the search process of 1NN is finished.

4.2 DIST_BF Optimization Algorithm Based on Distance Pruning Strategy

First we give the non-leaf nodes in the SIR-tree about the query q and thus determine whether the object in that node can be a candidate or not. Then we propose an upper limit on the distance between the query q. If the algorithm traverses a node whose distance is greater than this upper limit, the node and its subtree can be pruned and the algorithm terminates early because none of the objects in the node can be called candidates. These two pruning strategies are formally proposed below.

Lemma 1: Let the maximum priority queue *Result* hold the top-k space-text objects retrieved by the algorithm, objects $o \in Result$ in the queue according to the ranking score $rank_q(o)$ in descending order. Given a *SKSK* query and a non-leaf node in the SIR-tree R, the R on the query q the upper bound on the ranking score is $rank_q(R)$, if $rank_q(R) \le rank_q(o_k)$, all objects in R objects in can be pruned.

Proof: Definition 3 proves that given a query q and a non-leaf node R that for R all entries in e have $rank_q(e) \le rank_q(R)\}$. If $rank_q(R) \le rank_q(o_k)$, then R all the entries in e of the ranking function are less than $rank_q(o_k)$. Therefore, the node R and its subtree can be trimmed.

Lemma 2: Given a query q and a maximum priority queue *Result that* stores all candidate objects, if the algorithm encounters a query whose distance from it to q distance is not less than $\lambda = \frac{\Phi(o_k)^{\frac{1}{}} - 1}{\alpha}$ object or node, the algorithm stops.

Proof: If a non-leaf node R has a ranking score of $rank_q(R) \le rank_q(o_k)$, the node should be cropped according to Lemma 1. $rank_q(R) = \varepsilon * dp_q(R) + \beta * tr_q(R) + (1 - \varepsilon - \beta) * sd_q(R) \le rank_q(o_k)$, assuming that both textual relevance and social relevance are 1 (maximum), the $rank_q(R) = \varepsilon * dp_q(R) + (1 - \varepsilon) \le rank_q(o_k)$, i.e. $dp_q(R) \le 1 + \frac{rank_q(o_k) - 1}{\varepsilon}$, and because $dp_q(p) = 1 - \frac{dist(q,p)}{D_{max}}$, so when $dist(q, p)$ is greater than $\lambda = \frac{rank_q(o_k) - 1}{\varepsilon} * D_{max}$ when the node and its subtree can be trimmed. Because the queue that holds the candidate objects is a minimum priority queue, i.e., the objects are listed in ascending order according to their distance to the query q Euclidean distance to the query, when the distance is greater than λ the object in the queue is also greater than λ Therefore, the algorithm can be terminated early.

We apply Lemmas 1 and 2 to the BF algorithm proposed above, and the algorithm pseudo-code is as follows. (Lines 2–3) Initialize a minimum priority queue Queue used to hold the candidate objects, which are ordered according to the distance between the object location and the query point $dist(q, p)$ in ascending order. Initialize another max-priority queue Result that holds the top-k objects, which are sorted in $rank_q(.)$ ascending order. (line 5) The root node of the SIR-tree index and its key $dist(q, index.RootNode)$ into Queue in the SIR-tree index. (Lines 6–7) When Queue there are still candidate objects in the SIR-tree index, the loop will return the first object in the priority queue. e The first object in the priority queue is returned. (line 8) If $dist(q, e) < \lambda$, according to Lemma 2, the first object in the Queue all the objects after the object e all objects after can not be called candidates, the algorithm ends and returns Result. (Lines 9–14) If e is a node, iterate through e each child node in e if $rank_q(e)$ is greater than the value of Result the last spatial object in o_k the value of the ranking function $rank_q(o_k)$ then the e is inserted into Queue into the (Lines 15–19) If e is a space object and $rank_q(e) > rank_q(o_k)$, then the e is removed from Queue is removed, and a breadth-first search is used to determine the social relevance of the visited node $sd_q(e)$. Calculate the e The ranking function of $rank_q(e)$ that will $rank_q(e)$ as e the key value is inserted into the priority queue Result in the queue and update the λ the value of the (Lines 20–21) If the Result the number of space objects in k is reached, the algorithm ends.

```
Algorithm 3: DIST_BF Optimization Algorithm
1:  function SIRTQ(Query q, Integer k, SIR-tree index)
2:      Queue ← a new min-priority queue
3:      Result ← a new max-priority queue
4:      λ ← ∞
5:      Queue.Enqueue(index.RootNode, dist (q, index.RootNode));
6:      while not Queue.IsEmpty() do
7:          Entry e ← Queue.Dequeue();
8:          if dist(q, e) < λ then
9:              if e is a non-leafnode then
10:                 for each entry e' in e do
11:                     if rank_a (o_k) < rank_q (e') then
12:                         Queue.Enqueue (e', dist (q, e'))
13:                     end if
14:                 end for
15:             else if e refers to an object then
16:                 if rank_q (o_k) < rank_q(e) then
17:                     delete e from Queue
18:                     Compute its social relevance sd_q(e) using BFS;
19:                     Result.Enqueue (e, rank_q(e) )and update λ
20:                     if k objects have been found then return the top-k Result;
21:                 end if
22:             end if
23:         end if
24:         else break
25:         end if
26:         return Result;
27:     end while
28: end function
```

4.3 An X-HOP_BF Optimization Algorithm Based on Social Distance Pruning Strategy

The literature [5] proposes an X-HOP algorithm applied to $SKSK$ queries, which prunes social relevance based on a preferential traversal algorithm, and researchers argue that it is difficult to influence the preferences of two users on each other when they are too far apart in the undirected graph of a social network. The function to calculate social relevance $sd_q(.)$ The parameter in $dist_s(u_q, u)$ denote the query user u_q and user u in the undirected graph, and the X-HOP algorithm completes the pruning of social relevance by restricting the distance between two users in the undirected graph $(dist_s(u_q, u) \leq x)$. In the X-HOP algorithm $sd_q(.)$ the function is updated as

$$sd'_q(p) = \sum_{u \in p.M \wedge dist_s(u_q,u) \leq x} \alpha^{dist_s(u_q,u)} \tag{3}$$

We apply this pruning strategy to the distance-based BF algorithm and propose the X-HOP_BF algorithm. This query algorithm performs top-k queries, except that the line 18 of the Algorithm 3 pseudo-code used by the $sd_q(.)$ function in the pseudo-code of Algorithm is updated to $sd_q'(.)$, the idea is exactly the same as Algorithm 3.

5 Experiment

Environment: The experimental environment is an intel® Core™ i5-6300U 2.40 GHz CPU with 8 GB RAM, using windows 10 operating system and C++ language implementation.

Dataset: Brightkite[1] and Gowalla[2] are both location-based online social networks where users can share their location through check-ins and social networks are collected from public APIs.

Table 2. Specific information on the datasets Brightkite and Gowalla

	Brightkite	Gowalla
Data collection time	April 2008–October 2010	February 2009–October 2010
Number of users	58228	196591
Number of social connections	214078	950327
Number of check-in messages	4491143	6442890
Number of check-in positions	772996	1280969
Check-in data logging	User id, check-in time, latitude and longitude of check-in location, location id	User id, check-in time, latitude and longitude of check-in location, location id

---

[1] http://snap.stanford.edu/data/loc-Brightkite.html.

[2] http://snap.stanford.edu/data/loc-gowalla.html.

Since the Brightkite dataset and the Gowalla dataset only provide the check-in location and do not provide a textual description of the check-in location, we also need to add a descriptive text for each check-in location. For the description text dataset we used the Datafiniti_Hotel_Reviews[3] dataset, which is sourced from the Datafiniti Business Database and contains data from 10,000 user reviews of 1670 hotels between January 2018 and September 2018. The content of the dataset is shown in Table 2. We set the length of each text description to 20 words, which consists of the hotel name (name) and the review text (reviews.txt). The search text length is set to 5 words, and in this experiment we filter out the titles of reviews (reviews. title) with length greater than or equal to 5, and use the first 5 words of the title as the search text.

To test the pruning effect of the optimization algorithms, we apply the *BF* algorithm, distance-based *BF* algorithm, X-HOP algorithm, and X-HOP_BF algorithm, simultaneously on two datasets to measure the time taken to search for top-k results. To measure the similarity of the two algorithms returning top-k, we used the Kendall distance (K_{min}) [7] to measure it, which responds to the percentage of pairs in the two top-k lists that are arranged in opposite order.K_{min} The smaller it is, the more similar the two top-k lists are. The values of the parameters set in the experiments were derived empirically.

5.1   Exploring the Impact Parameters α on the Query Efficiency

Figure 6 and Fig. 7 show the time taken by the X-HOP algorithm, X-HOP_BF algorithm, and X-HOP_BF algorithm in varying the impact parameters α the time taken by the *BF* algorithm, distance-based *BF* algorithm, X-HOP algorithm, and X-HOP_BF algorithm to perform top-k search on the datasets Gowalla and Brightkite when changing the influence parameters. In the experiments, let $k = 20$, $\varepsilon = 0.5$. $\beta = 0.3$. The running time of the four algorithms increases as α increases with the increase of The smaller α indicates that social relevance has a small impact on the ranking function, and larger α indicates that social relevance has a greater impact on the ranking function. The reason for this experimental result may be that the larger α expansion of the search scope, resulting in an increase in search time. From the figure, it can be learned that the efficiency of the optimized *BF* algorithm is obviously due to the original *BF* algorithm from the perspective of running time, and the time of all three optimized algorithms is 1/2 of the original *BF* algorithm. it can be seen from Fig. 6 that the running efficiency of the X-HOP_BF algorithm is slightly better than that of the distance-based *BF* algorithm and the X-HOP algorithm, which is because the algorithm prunes from the perspective of both distance and social relevance, and greatly reduces the search space.

---

[3] https://www.kaggle.com/datafiniti/hotel-reviews?select=Datafiniti_Hotel_Reviews.csv.

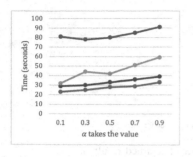
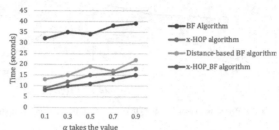

Fig. 6. Gowalla data for exploring the impact parameter α

Fig. 7. Brightkite data for Exploring the impact parameter α.

5.2 Exploring the Number of Returned Results k on the Query Efficiency

Figure 8 and Fig. 9 show the time taken by the *BF* algorithm, distance-based *BF* algorithm, X-HOP algorithm, and X-HOP_BF algorithm in changing the number of returned results k *BF* algorithm, distance-based *BF* algorithm, X-HOP algorithm, and X-HOP_BF algorithm when changing the number of results returned, the time taken by the *BF* algorithm, distance-based *BF* algorithm, and X-HOP_BF algorithm to obtain top-k results for the data sets Gowalla and Brightkite and K_{min}. As we expected, the running times of the four algorithms increased with k increasing slightly. Because the more objects returned, the higher the computational overhead required. From the running time perspective, the three optimized algorithms run significantly more efficiently than the unoptimized *BF* algorithm. Among them, the X-HOP algorithm outperforms the distance-based *BF* algorithm in both datasets due to the greatly reduced complexity of the breadth-first traversal algorithm. The X-HOP_BF algorithm performs the best among the four query algorithms because it prunes from the distance between two objects, and the social relevance between two users simultaneously. We use the top-k objects returned by the *BF* algorithm as the benchmark, compare the top-k objects returned by the other 3 algorithms with them, and calculate the corresponding K_{min} values. From Fig. 9, we can see that the 3 algorithms' K_{min} values decrease as k This indicates that the pruning algorithm only affects the ranking of a small number of objects, and the algorithms return more similar results when the number of results returned is requested.

This Section demonstrates the efficiency and scalability of the proposed algorithms by running the *BF* algorithm, distance-based *BF* algorithm, X-HOP algorithm, and X-HOP_BF algorithm on real data sets Brightkite and Gowalla. Firstly, the parameters are explored α on the *SKSK* query, it is known from the experimental results that the query algorithm running time increases with α increases, and the runtime of the three algorithms based on the optimization is significantly shorter. Among them, the X-HOP_BF algorithm is the most efficient among the four query algorithms. Then explore the effect of the number of returned results k on the *SKSK* query, with k increases, the running time of the four algorithms increases significantly, but the set of results returned by the four algorithms tends to be similar. These two experiments demonstrate the optimization effect of the distance-based pruning strategy and the social relationship-based pruning strategy for *SKSK* queries, as well as the consistency of the search results while improving the running speed.

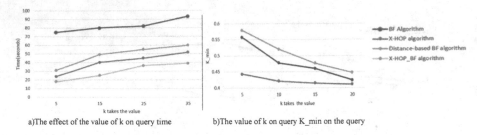

a)The effect of the value of k on query time   b)The value of k on query K_min on the query

Fig. 8. Gowalla data for exploring the number of returned results k

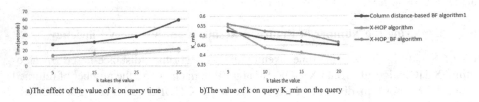

a)The effect of the value of k on query time   b)The value of k on query K_min on the query

Fig. 9. Brightkite data for exploring the number of returned results k

6 Conclusion

In order to extend the application scope of spatial keyword query and better serve users' personalized query requirements, this paper investigates the social-aware top-k keyword query, which enriches the semantics of spatial keyword query by introducing a new social relevance attribute. In terms of index structure, we adopt a hybrid index structure, SIR-Tree, which integrates social information into the IR-tree and calculates the social relevance of spatial-text objects to query users through their social networks. Based on the SIR-tree, we propose a top-k query algorithm *BF* based on the idea of prioritized traversal. since the query needs to calculate spatial distance, text similarity, and social relevance simultaneously, which generates a large amount of computation during the search process, we propose a corresponding pruning strategy to optimize the *BF* algorithm. The final experimental results show that the optimized *BF* algorithm has good performance compared with the original *BF* algorithm.

Acknowledgment. We thank the reviewers from NPC 2021 for their valuable feedback and guidance. Zhixin Zeng is the corresponding authors of the paper. This work is supported by the Natural Science Foundation of Guangdong, China under 2021A1515011755.

References

1. Cui, N., Li, J., Yang, X., et al.: When geo-text meets security: privacy-preserving boolean spatial keyword queries. In: 2019 IEEE 35th International Conference on Data Engineering (ICDE), pp. 1046–1057. IEEE (2019)

2. Yang, J., Zhang, Y., Zhou, X., et al.: A hierarchical framework for top-k location-aware error-tolerant keyword search. In: 2019 IEEE 35th International Conference on Data Engineering (ICDE), pp. 986–997. IEEE (2019)
3. Xu, H., Gu, Y., Sun, Y., et al.: Efficient processing of moving collective spatial keyword queries. VLDB J. **29**(4), 841–865 (2020)
4. Dong, Y., Xiao, C., Chen, H., et al.: Continuous top-k spatial-keyword search on dynamic objects. VLDB J. **30**(2), 141–161 (2021)
5. Li, J., Wang, S., Qin, S., Li, X., Wang, S. (eds.): ADMA 2019. LNCS (LNAI), vol. 11888. Springer, Cham (2019). https://doi.org/10.1007/978-3-030-35231-8
6. Chen, L., Cong, G., Jensen, C.S., et al.: Spatial keyword query processing: an experimental evaluation. Proc. VLDB Endowment **6**(3), 217–228 (2013)
7 Fagin, R., Kumar, R., Sivakumar, D.: Comparing top k lists. SIAM J. Discrete. Math. **17**(1),134–160 (2003)

System Software and Resource Management

A Novel Task-Allocation Framework Based on Decision-Tree Classification Algorithm in MEC

Wenwen Liu, Zhaoyang Yu, Meng Yan, Gang Wang$^{(\boxtimes)}$, and Xiaoguang Liu$^{(\boxtimes)}$

TJ Key Lab of NDST, College of Computer Science, Nankai University,
Tianjin, China
{liuww,yuzz,yanm,wgzwp,liuxg}@nbjl.nankai.edu.cn

Abstract. Mobile-edge computing (MEC) has emerged as a promising paradigm to extend the cloud computing tasks to the edge mobile devices for improving the quality of service. This paradigm addresses the problems in cloud computing architecture by enabling lower latency, higher bandwidth, and better privacy and security. Previous studies of task allocation in MEC systems generally only consider the single influence factor, such as the distance between the mobile device and cloudlets, to select the suitable cloudlets for tasks. However, there are various types of tasks with complex individual requirements about which we should consider, such as transmission bandwidth, computing capacity and storage capacity of cloudlets, and so on. In this paper, we propose an on-demand and service oriented task-allocation framework based on machine learning technology, called **Edgant**. It classifies tasks into three types using a decision-tree model according to the tasks' characteristics including requirement on server resources and user's requirements. For each type, we provide a selection strategy to allocate task to the most suitable cloudlet based on characteristics of the task and the current state of cloudlets. Simulated evaluations demonstrate that Edgant achieves lower latency and better accuracy compared to other task allocation methods, as well as high reliability under high mobility of the mobile devices.

Keywords: Mobile edge computing · Decision-tree classification algorithm · Task-allocation framework · Lower latency · More accurate · More reliability transmission

1 Introduction

The emerging Internet of Things (IoT) and Fifth Generation (5G) mobile network enable many promising applications. For example, autonomous vehicles [11]

This work is partially supported by Science and Technology Development Plan of Tianjin (18ZXZNGX00140, 18ZXZNGX00200, 20JCZDJC00610); National Science Foundation of China (U1833114, 61872201).

make decisions by collecting videos from cameras and data via sensors. Computational tasks are often off-loaded to cloud due to vehicles' limited resources, but this may result in a high latency, which is unacceptable, even dangerous, in case of accidents. Edge computing [2,12,15] has been proposed to address previous issues, in which small servers popularly called cloudlets [12,18] are placed at the edge of the Internet, in close to end users, thus extends the cloud–device architecture to the cloud–edge–device architecture. A MEC system uses some task-allocation algorithm to discover and select a suitable cloudlet for each task from a mobile device. However, the current task-allocation algorithms generally only consider a single influence factor for various types of tasks, such as the distance between devices and cloudlets [18] and neglects other factors: transmission bandwidth, computing capacity, storage capacity, and stored data contents of cloudlets. This will lead to unsatisfied performance which may be dangerous in some mission-critical applications, such as autonomous-vehicle.

To accommodate various application scenarios, we present a novel on-demand task-allocation method called **Edgant**. Compared with the cloudlet-discovery mechanism that only considers the single factor such as distance, it takes various factors into account, carefully divides the tasks into three types through a decision tree model and adopts different strategies to schedule them. The main contributions of this paper are as follows.

- We propose a novel on-demand task-allocation framework called Edgant which divides incoming tasks into three types using decision trees.
- We design customized task-allocation strategy for each type of tasks to select the suitable cloudlet for them.
- We evaluate Edgant with various tasks and show its advantages over two traditional frameworks in terms of latency, accuracy, and reliability.

The rest of the paper is organized as follows. Section 2 provides related work and Sect. 3 describes the design of Edgant in detail. The simulation experiments and analysis are displayed in Sect. 4. The application scenarios of Edgant are discussed in Sect. 5 and Sect. 6 concludes this paper.

2   Related Work

Task allocation has been extensively studied in cloud computing community. Wang et al. [22] proposed the Task Share Fairness (TSF) policy for jobs with diverse demands better off sharing the data center resources. An altruistic, long-term approach was proposed in [6], called CARBYNE. It allocates resources to jobs without impacting their own completion time. Tang et al. proposed the Long-Term Resource Fairness (LTRF) method to solve the aforementioned problems both in single-level and hierarchical resource allocations for user to submit different meaningful workloads [21]. For heterogeneous environments, Khamse-Ashari et al. [9] developed Per-Server Dominant-Share Fairness (PS-DSF) mechanism. It compares and weighs the allocated resources by considering heterogeneity to achieve desirable properties. Then Khamse-Ashari et al. [8] proposed a specific class of utility functions that trades off between efficiency and fairness in a distributed fashion.

For task allocation in edge computing, there have also been some studies. In [18], the authors proposed an algorithm that selects the cloudlet that has the shortest network distance from the mobile device. In [14], the authors proposed a power and latency aware optimum cloudlet selection strategy. Hend Ben Saad et al. [16] proposed a cloudlet selection strategy which ensures both good network connection and fast execution of the application. Zhang et al. [25] designed a load-aware resource allocation and task scheduling (LA-RATS) algorithm to deal with both delay-tolerant and delay-sensitive mobile applications.

In particular, some work focuses on task allocation in intelligent automatic controlling. The obvious challenges are ubiquitous wireless networks connections and high quality of service for numerous vehicles. To tackle these issues, Zhang et al. [26] proposed a new Vehicular Edge Computing (VEC) framework to address the tasks offloading difficulty caused by high mobility of the vehicles. Another study [23] on this problem proposed a location-based offloading scheme to reach good trade-off between the task completion latency and the resource utilization of wireless network systems. Based on this study, Huang et al. [7] proposed three data centers deployment optimization schemes to optimize data collection and code dissemination tasks.

Existing studies on task-allocation strategy either selects the closest cloudlet or uses a single strategy for various types of tasks, which may not consider other factors that affect quality of service and meet various requirements of different types of tasks. In this paper, we propose an on-demand task-allocation framework that called Edgant. It could dynamically achieve the suitable cloudlet-selection solution for various task types to encourage cloudlet operators.

3   The Edgant Framework

3.1   The Architecture Overview

As Fig. 1 shows, a MEC system based on Edgant involves three layers: 1) the *mobile-device layer* contains end users, mobile devices and sensors, 2) the *cloudlet layer* contains common cloudlets (edge servers in edge computing) and a cloudlet manager, 3) and the *cloud layer*, it serves as a monitor of the whole system. The executing procedure of a task is as follows.

1. The mobile device sends a task along with its requirements for interaction frequency, CPU capacity, memory size and data content to the cloudlet manager.
2. The cloudlet manager categorizes the received task into one of three types (see Sect. 3.2) and selects the most suitable cloudlet to execute the task based on the corresponding task-allocation strategy (see Sect. 3.3). It also updates the state of the cloudlet.
3. The selected cloudlet executes the task and sends the result back to the mobile device.

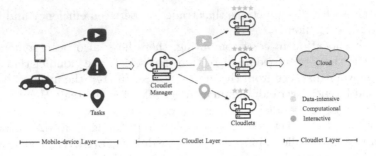

Fig. 1. The overview of the Edgant.

3.2 Task-Classification Model

As mentioned in [19], in MEC systems, the tasks can be divided into three types, interactive task, computational task and data-intensive task, according to various requirements tabulated in Table 1. In this section, we will introduce how to use the *Decision Tree* algorithm to implement task classification.

Table 1. The types of off-loading tasks.

Task types	Features	Influencing factors
Interactive task	Frequent interactions with terminal devices	Bandwidth, distance etc.
Computational task	Massive calculation in cloudlets	Hardware resources: CPU, memory etc.
Data-intensive task	Various kinds of data need to be accessed	The data content stored in cloudlets

In our work, we use the popular C4.5 algorithm [13] to build our task classification models. In the beginning, we only have the root node containing all of the training samples. Then we select the attribute with the highest information gain to split the root node, that is, create its two children. We recursively perform split until the stop criteria holds. Now we have a decision tree for task classification. When we determine the type of a task, we start from the root node and go down to the children whose branch matches the corresponding attribute value of the task. We repeat this step until get to a leaf, and then the task will be assigned to the same type as the majority in the leaf.

In our experiments, we use the training set containing 216 tasks to establish the decision tree. Each task has quadruple attributes as the input:

- *P.freq* measures the frequency of the required interactions between the cloudlet and the mobile device during task execution.
- *P.cpu* denotes the amount of CPU resource required to perform the task.

- *P.mem* is the amount of memory resource required to perform the task.
- *P.data* specifies what data is required to perform the task.

Algorithm 1 summarizes the task classification procedure based on the constructed decision tree, where *CPUT* and *MemT* are predefined thresholds for CPU and memory resources respectively. We test the classifier on other 100 tasks, and the results that our decision tree achieves a classification accuracy rate of 89.23%.

Algorithm 1. Task Classification Algorithm Base on Decision-Tree

---

Input: The quadruple of task parameters $P = \{freq, cpu, mem, data\}$
Output: The type of task $t \in \{Interactive, Computational\ task, Data\text{-}intensive, Cloud, Local\}$
1: **if** $P.freq \geq FreqT$ **then**
2:     **return** *Interactive*
3: **end if**
4: $S_{cpu} \leftarrow \{C \mid P.cpu \leq C.cpu\}$
5: $S_{mem} \leftarrow \{C \mid P.mem \leq C.mem\}$
6: **if** $S_{cpu} \cap S_{mem} = \varnothing$ **then**
7:     **return** *Cloud*
8: **end if**
9: **if** $P.cpu > CPUT$ **and** $P.mem > MemT$ **then**
10:     **return** *Computational*
11: **end if**
12: **if** $\mathbf{type}(P.data) = special$ **then**
13:     **return** *Data-intensive*
14: **else**
15:     **return** *Local*
16: **end if**

3.3   Task-Allocation Strategies

In this section, we will detail how to assign each type of tasks to cloudlets.

Interactive Task. For an interactive task, such as the path planning task in automatic driving, a lot of communication between the mobile device and the cloudlet should be performed. In [24], an equation about Received Signal Strength Indication (RSSI) is given by,

$$\|RSSI\| = A + 10n \cdot \lg d. \tag{1}$$

We can see that the transmission distance d is an important factor affecting the signal strength (A and n are is dictated by the environment and it is hard to change them to improve RSSI) and hence an important influence factor of network transmission performance and stability. So we assign an interactive task to the cloudlet with the shortest **Network Distance** from it.

We denote the mobile network through a weighted undirected graph G with the set of vertices denoting the mobile devices and the cloudlets, and the weight of the edge between two vertices representing the Euclidean distance between them. Given a certain mobile device m and a set of reachable cloudlets $C = \{c_1, c_2, \ldots, c_k\}$, we define the network distance D_{m,c_i} between m and c_i as the average path length [4]. Then the most suitable cloudlet for m is the one obtains the shortest network distance $\arg\min_i D_{m,c_i}$.

Algorithm 2. Computational Task-Allocation based on Dominant Resource

---

Input: The set of cloudlet severs, $S = \{s_1, s_2, s_3, \cdots, s_j\}$
    The amount of hardware resource for current task P.
Output: A suitable cloudlet c.
 1: $R \leftarrow \varnothing$
 2: **for all** $s_i \in S$ **do**
 3:    **if** $s_i.avail\_cpu \geq P.cpu$ **and** $s_i.avail\_mem \geq P.mem$ **then**
 4:        $r_c \leftarrow \frac{P.cpu}{s_i.avail\_cpu}$
 5:        $r_m \leftarrow \frac{P.mem}{s_i.avail\_mem}$
 6:        Label the resource with the larger ratio as the dominant resource of s_i
 7:        $R.\mathbf{add}(\max(r_c, r_m))$
 8:    **end if**
 9: **end for**
10: sort R in descending order
11: **return** s_i with the highest ratio in R.

Computational Task. A computational task, such as 3D object detection tasks and audio or video processing tasks in autonomous vehicle, involves a great deal of calculation and requires server with higher performance. So when assign this type of tasks, we mainly consider unused CPU and memory capacity of cloudlets. We develop a greedy algorithm to select appropriate cloudlet for a given computational task, which is summarized in Algorithm 2. For every cloudlet meeting both CPU and memory requirements of the task, we calculate two ratios of the required resource of the task to the available resource of the cloudlet (Line 4 and Line 5). Then we compare the magnitude of the two ratios, and label the resource with the larger ratio as the dominant resource of the current cloudlet. After that, we sort the candidates in descending order of the dominant resource ratios and select the cloudlet with the highest ratio as the most suitable one for the current task. This best-fit like algorithm ensures that the selected cloudlet could perform the task efficiently, and manages to minimize resource fragmentation.

Data-Intensive Task. For the data-intensive tasks, specific data needs to be accessed, and the cloudlet manager assigns the cloudlet that contains the relevant data. For example, an user in vehicular networks may request map to plan the driving route and the cloudlet storing the map will be chosen to perform this kind of map inquiry tasks. For the entertainment systems in vehicles, the cloudlet manager will choose the cloudlets that storing videos, pictures and music. The information about which cloudlet stores which data is recorded in the cloudlet directory maintained by the cloudlet manager.

4 Experiments

Table 2. Summary of the tasks used in evaluation.

Application	Task type	Description
Realtime positioning [20]	Interactive task	1,000 records of accelerometer each of 7 classes
Feature extraction [10]	Computational task	6000 examples of each of 10 classes
Object detection [3]	Data-intensive task	640 * 360 images with 3D object

4.1 Experiments Setup

We use the widely-used cloudlet-discovery project [18] to implement cloudlets and modify it to implement our cloudlet manager. The Network Time Protocol (NTP) is used to synchronize time for mobile devices, cloudlets and cloud servers to measure task latency of tasks. The bandwidth is set to 8 Mbps according to Internet connection speeds in different countries in [1]. We use 316 autonomous vehicle tasks [5], and 216 of them for training and the rest for test. They cover all the three task types. Real-time positioning belongs to interactive task, feature extraction is a computational task, and object detection is data-intensive. The detailed information is summarized in Table 2.

We compare Edgant with two baselines, the Distance-Only framework [17] and the Cloud-Only framework.

– **Cloud-Only framework:** The user directly transmits the tasks to the remote cloud to run. There is no edge servers in the system.
– **Distance-Only framework:** The task-allocation algorithm selects the closest edge server to run the tasks. Although the shortest distance reduces the distance and therefore the data transmission time successfully, it does not consider whether the shortest distance server is suitable for task execution or not.

The configuration of whatever is shown in Table 3. The Distance-Only framework doesn't use the cloudlet manager, and the closest cloudlet to the current mobile device is always selected. The Cloud-Only framework doesn't involves the cloudlets and the cloudlet manager, and all the tasks are directly sent to the cloud server.

Table 3. Configuration of the evaluation platform.

Equipment	Frequency of CPU	Core of processor	Memory
Cloudlet manager	2.4 GHz	2 cores	4 G
Common cloudlet 1	2.4 GHz	2 cores	4 G
Common cloudlet 2	2.4 GHz	4 cores	8 G
Common cloudlet 3	2 GHz	2 cores	2 G
Cloud server	3.2 GHz	6 cores	16 G
Mobile device	1.2 GHz	2 cores	1 G

4.2 Latency Evaluation

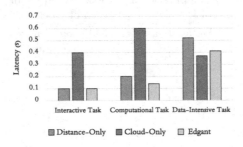

Fig. 2. The comparison of latency of three types of tasks

Latency is an important metric in the real-time scenarios. Here latency consists of the transmission time and the execution time. The transmission time represents the time of mobile device sending data to the cloudlet (or the cloud server) and getting results from it, and also includes the cloudlet-selection time in Edgant and the Distance-Only framework. The execution time is the time of executing the task in the cloudlet (or the cloud server).

Figure 2 shows the overall latency of different types of tasks. It can be seen that when performing interactive tasks and computational tasks, Edgant achieves the lowest latency and the latency of Cloud-Only is rather high. For data-intensive task, although Edgant's latency is a little higher than that in the Cloud-Only framework, it is still acceptable and is lower than that in the Distance-Only framework.

To better understand the latency of different types of tasks in the three frameworks, we break down the latency into transmission time and execution time as Fig. 3 shows.

We can see that, for the interactive task, the transmission time dominates the overall latency because of frequent interactions between mobile devices and cloudlets/cloud servers. Since Edgant selects cloudlets according the network distance which is the main influence factor on transmission performance, its

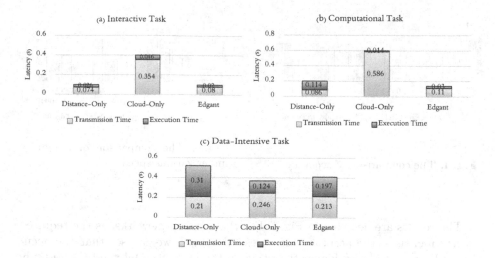

Fig. 3. The transmission time & execution time of (a) interactive task (b) computational task and (c) data-intensive task.

transmission time is close to that of the Distance-Only framework. In contrast, in the Cloud-Only framework, the cloud server is far away from the mobile device which results in very high transmission time.

For the computational tasks, the cloudlet manager in Edgant will allocate a cloudlet with sufficient hardware resources for the mobile device. But the selected cloudlet is not necessarily the one closest to the mobile device. We can see from Fig. 3(b) that, the transmission time of Edgant is actually longer than that in the Distance-Only framework, but much shorter than that in the Cloud-Only framework. On the other hand, compared with the Distance-Only framework, Edgant provides much shorter execution time and obtains lower overall latency.

For data-intensive tasks, since the data transfer is not frequent, the transmission time in the Cloud-only framework is close to those in the other frameworks, and it provides the lowest overall latency because of its powerful computation capacity. Edgant selects the cloudlet storing related data, so it takes less time to perform the task compared with the Distance-Only framework.

In a word, since Edgant takes multiple factors into consideration, it achieves more balanced performance compared with the baselines.

4.3   Accuracy Evaluation

Only timely response is not enough for the mobile applications and the accuracy also need to be taken into consideration. For example, a minor error for the feedback of the road conditions in the automatic driving scenario may lead to serious traffic accidents. We carry out the computational task to measure the accuracy of the three frameworks under various latency requirements (limits). Task execution is stopped when a given latency limit is reached. We compare the accuracy of image recognition, that is, the probability of correct image recognition.

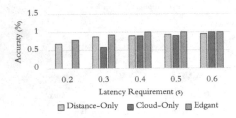
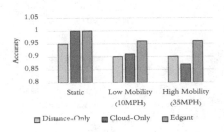

Fig. 4. The comparison of accuracy

Fig. 5. The comparison of reliability under various speeds

The results are depicted in Fig. 4. For the three frameworks, as the required latency increases, the accuracy increases. Moreover, we can see that the accuracy of Edgant is always higher than that in the Distance-Only and Cloud-Only frameworks. This is because Edgant allocates cloudlets with sufficient computing resources to tasks, which is helpful for obtaining higher accuracy. On the other hand, when the require latency is 0.2 s, the Cloud-Only framework even cannot return any result in time. Moreover, at the latency requirement of 0.6 s, the accuracy of Edgant is about 99%, which is roughly the same as the accuracy of the Cloud-Only framework when given large enough latency. The results indicates that Edgant can be applied well to the application scenarios require for high-precision results.

4.4 The Reliability in a Mobile Environment

In this subsection, we explore the reliability, which includes the accuracy which is same as Sect. 4.3 under the 0.6 s latency requirement in three frameworks while meeting the mobility demand. When the wireless mobile devices moving from one area to another, the frameworks should ensure that their communication connections move continuously and the services they enjoyed are not affected. The reliability is measured by accuracy under three states of the mobile device including static, low mobility and high mobility. The speed of mobile device at low mobility is 10 mi per hour (MPH) and the speed at high mobility is 35 MPH. The result is displayed in Fig. 5 and the corresponding observations and analysis are as follows.

First, the accuracy decreases as the speed of the mobile device increases. The reason is obvious. On the one hand, the higher speed of mobile device makes the network unstable. On the other hand, the quality of the network link may decrease during the switch between different networks in the mobile environment.

Second, for the three frameworks, we can see that the reliability of Edgant framework is higher than the Distance-Only framework and Cloud-Only framework at various moving speed. Compared with Distance-Only framework, the Edgant framework allocates cloudlets with sufficient computing resources to the task While ensuring the quality of network transmission in the mobile process, however the Distance-Only framework may have the problem of insufficient computing resources, therefore the Edgant's result is higher than Distance-Only framework. On the other hand, for the Cloud-Only framework, no matter where the mobile terminal moves, it is connected with the cloud center fixed at a point. So with the increase of moving speed, the Cloud-Only framework can't guarantee the quality of network connection. And Edgant will select the edge server of network transmission quality in the process of moving. Furthermore, at the latency requirement of 0.6 s, the reliability of Edgant framework is able to achieve not be lower than 96%, which maintains the highest reliability at different moving speeds. The results demonstrate a better reliability performance of Edgant framework.

5 Conclusions

In this paper, we present Edgant framework to ensure that tasks with various demands run efficiently. The Edgant framework divided tasks into three types using decision tree classification algorithm according to the attributes and demands of tasks in MEC system. Then we provided a task-allocation strategy for each type of task to ensure efficient, accurate and timely running of tasks with different demands and characteristics. The evaluations were presented to illustrate the effectiveness of the Edgant framework and demonstrate the superior performance over the existing traditional task-allocation frameworks.

In the future, some other issues could be considered. For example, how to predict server resource consumption in a dynamic environment for tasks to get more convenient classification of tasks.

References

1. List of countries by internet connection speeds. https://en.wikipedia.org/wiki/List_of_countries_by_Internet_connection_speeds. Accessed May 2020
2. Ahmed, A., Ahmed, E.: A survey on mobile edge computing. In: International Conference on Intelligent Systems and Control (2016)
3. Collet, A., Martinez, M., Srinivasa, S.S.: The moped framework: object recognition and pose estimation for manipulation. Int. J. Robot. Res. 30(10), 1284–1306 (2011)
4. Fronczak, A., Fronczak, P., Hołyst, J.A.: Average path length in random networks. Phys. Rev. E: Stat., Nonlin, Soft Matter Phys. 70(2), 056110 (2004)
5. Geiger, A., Lenz, P., Stiller, C., Urtasun, R.: Vision meets robotics: the KITTI dataset. Int. J. Robot. Res. (IJRR) 32, 1231–1237 (2013)
6. Grandl, R., Chowdhury, M., Akella, A., Ananthanarayanan, G.: Altruistic scheduling in multi-resource clusters. In: USENIX OSDI (November 2016)
7. Huang, B., Liu, W., Wang, T., Li, X., Song, H., Liu, A.: Deployment optimization of data centers in vehicular networks. IEEE Access 7, 20644–20663 (2019)
8. Khamse-Ashari, J., Lambadaris, I., Kesidis, G., Urgaonkar, B., Zhao, Y.: An efficient and fair multi-resource allocation mechanism for heterogeneous servers. IEEE Trans. Parallel Distrib. Syst. 29, 2686–2699 (2018)

9. Khamse-Ashari, J., Lambadaris, I., Kesidis, G., Urgaonkar, B., Zhao, Y.: Per-server dominant-share fairness (PS-DSF): a multi-resource fair allocation mechanism for heterogeneous servers. In: 2017 IEEE International Conference on Communications (ICC), pp. 1–7. IEEE (2017)
10. Krizhevsky, A., Hinton, G.: Learning multiple layers of features from tiny images. Technical report. Citeseer (2009)
11. Liu, S., Tang, J., Zhang, Z., Gaudiot, J.L.: Computer architectures for autonomous driving. Computer **50**(8), 18–25 (2017)
12. Mach, P., Becvar, Z.: MEC: a survey on architecture and computation offloading. IEEE Commun. Surv. Tutor. **19**, 1628–1656 (2017)
13. Mantas, C.J., Abellán, J.: Credal-C4.5: decision tree based on imprecise probabilities to classify noisy data. Exp. Syst. Appl. **41**(10), 4625–4637 (2014)
14. Mukherjee, A., De, D., Roy, D.G.: A power and latency aware cloudlet selection strategy for multi-cloudlet environment. IEEE Trans. Cloud Comput. **7**, 141–154 (2016)
15. Mukherjee, M., Lei, S., Di, W.: Survey of fog computing: fundamental, network applications, and research challenges. IEEE Commun. Surv. Tutor. **20**, 1826–1857 (2018)
16. Saad, H.B., Kassar, M., Sethom, K.: Utility-based cloudlet selection in mobile cloud computing. In: 2016 Global Summit on Computer & Information Technology (GSCIT), pp. 91–96. IEEE (2016)
17. Satyanarayanan, M., Bahl, P., Caceres, R., Davies, N.: The case for VM-based cloudlets in mobile computing. IEEE Pervasive Comput. **8**(4), 14–23 (2009)
18. Satyanarayanan, M., Bahl, P., Cáceres, R., Davies, N.: The case for VM-based cloudlets in mobile computing. IEEE Pervasive Comput. **8**(4), 14–23 (2009)
19. Shi, W., Liu, F., Sun, H., Qingqi, P.: Edge Computing, pp. 133–136. Science Press, Beijing (2018)
20. Solenthaler, B., Pajarola, R.: Predictive-corrective incompressible SPH. ACM Trans. Graph. (TOG) **28**, 1–6 (2009)
21. Tang, S., Lee, B.S., He, B.: Fair resource allocation for data-intensive computing in the cloud. IEEE Trans. Serv. Comput. **11**(1), 20–33 (2018)
22. Wei, W., Li, B., Liang, B., Li, J.: Multi-resource fair sharing for datacenter jobs with placement constraints. In: International Conference for High Performance Computing (2016)
23. Yang, C., Liu, Y., Chen, X., Zhong, W., Xie, S.: Efficient mobility-aware task offloading for vehicular edge computing networks. IEEE Access **7**, 26652–26664 (2019)
24. Yuguo, Z.: Analysis of distance measurement based on RSSI. Chin. J. Sens. Actuators (2007)
25. Zhang, F., et al.: A load-aware resource allocation and task scheduling for the emerging cloudlet system. Fut. Gener. Comput. Syst. **87**, 438–456 (2018)
26. Zhang, K., Mao, Y., Leng, S., Maharjan, S., Vinel, A., Zhang, Y.: Contract-theoretic approach for delay constrained offloading in vehicular edge computing networks. Mob. Netw. Appl. **24**(3), 1003–1014 (2019)

QoS-Aware Scheduling for Cellular Networks Using Deep Reinforcement Learning

Jonathan Robert Malin, Gun Ko, and Won Woo Ro[✉]

Yonsei University, Seoul 03722, Republic of Korea
{jmalinza,gun.ko,wro}@yonsei.ac.kr

Abstract. This research presents a reinforcement learning (RL) approach that provides coarse-grained decisions to maximize QoS requirement satisfaction in mobile networks. Deep reinforcement learning has
demonstrated agents that can capture the dynamics of impossibly complex systems. At each scheduling interval, our RL agent provides a
scheduling policy that is suitable for optimal resource allocation in a
mobile network. By using a deep neural network to approximate the
action-value function (Q-function), scheduling decisions can be made
using an optimal policy. Utilising a 4G-LTE network simulator and
Pytorch, this research explores three scenarios of diverse traffic and
UE density. The implementation shows stable and effective performance
when compared to baseline static schedulers. Additionally, the RL agent
selects the optimal scheduler for both single and mixed traffic simulations. Being both scalable and cheap to compute, the implementation
offers a simple and effective method of radio resource management.

Keywords: Network scheduling · Deep reinforcement learning ·
Resource scheduling

1 Introduction

The 5th generation of mobile network technology (5G) is anticipated to revo-
lutionize society's access to information and move us closer to a fully mobile
and connected civilization. It is predicted that over 70% of the global popula-
tion will have a mobile connection by 2023, with 5G devices contributing over
10% [1]. Fuelled by the adoption of smart devices, consumers expect reliable
and high-speed network connections for data, video, and other content access.
The demand for data is forecast to increase seven-fold to over 77 exabytes per
month by the year 2022 [2]. Alarmingly, network management tools are usually
inherited from the previous generation [3]. In the past, these tools were well
maintained and updated to fit with the new network goals that they were being
adopted to. However, with the rapid and unprecedented scale-up of 5G network
requirements [4], this "adopt-and-improve" approach will not suffice.

© IFIP International Federation for Information Processing 2022
Published by Springer Nature Switzerland AG 2022
C. Cérin et al. (Eds.): NPC 2021, LNCS 13152, pp. 105–117, 2022.
https://doi.org/10.1007/978-3-030-93571-9_9

One such tool is the Radio Resource Management (RRM), which makes bandwidth/spectrum allocation decisions for users in the network. However, with a rising number of high-demand consumers, the average accessible network resource per user will be limited and the allocation decisions for those resources will be ever more complex. Therefore, RRM is poised to be a fundamental problem in 5G. In 5G's technological predecessor, 4G (also known as LTE) resource allocation is decided based on a single scheduler policy.

While this is valid in terms of the quality of service (QoS) requirements of consumers of mobile broadband, the usage patterns of 5G are slated to support a more varied collection of use cases and scenarios [4]. The rapid increase in users, diversity in traffic types, added use cases and the harsh QoS requirements imposed by data hungry content consumption all combine to present 5G as an impossibly complex system that is not suited to inherit the RRM tools from 4G.

Deep reinforcement learning offers solutions to innovate current RRM techniques. Applications in reinforcement learning can capture complex system dynamics and make optimal decisions for systems in autonomous driving [5], robotics [6], and AI gaming [7,8]. Reinforcement learning does not require any labelled data making training more flexible, thereby generalising decision algorithms and enabling constant learning.

This research proposes a QoS-guided scheduling method that makes coarse-grained decisions using a deep reinforcement learning agent. At the beginning of each transmission time interval (TTI), the agent selects a packet scheduler that defines a policy for resource allocation in the interval. Based on how a user's QoS requirements are being met, a user's channel quality index (CQI), and the number of users, the agent selects the scheduler that it believes will maximize the number of users with satisfied QoS requirements. Using the popular deep learning framework Pytorch [9], a deep Q-network (DQN) is implemented and trained using the 4G-LTE simulator LTE-Sim [10]. The experimental results demonstrate the ability of the DQN to provide fast and effective coarse scheduling decisions that are practical in terms of the timing requirements of current 4G networks.

2   Reinforcement Learning

Reinforcement learning (RL) categorises a set of learning methods focussed on learning to control a system, while maximizing a perceived reward in the long term. Compared to supervised learning, feedback on the learner's predictions is only partial and does not quantify the learner's error from the truth. The goal of reinforcement learning methods is to converge onto a sequence of decisions that ultimately maximizes a cumulative reward.

2.1   Value Functions

To understand the expected reward (or value) from a given state, the *state-value* function at some state s, following policy π, is defined as

$V^{\pi}(s) = \mathbb{E}[G_t \mid s_t = s] = \mathbb{E}[\mathcal{R}_{t+1} + \gamma V^{\pi}(s_{t+1}) \mid s_t = s]$. Where G_t is the cumulative discounted reward at time t and γ is called the discount factor and defines the importance between immediate and future rewards. Similarly, the *action-value* function for policy π, gives information about how rewarding any action a from a given state s will be, as $Q^{\pi}(s,a) = \mathbb{E}[r(s_t, a_t) + \gamma Q^{\pi}(s_{t+1}, a_{t+1}) \mid s_t = s, a_t = a]$.

The action-value function is also know as the *Q-function*. The decomposed version of both the value function and Q-function is known as the *Bellman Equation* [11]. The goal of the agent is to find the policy that will yield the most reward. In this research, we are interested in finding the optimal Q-function, Q^*. The optimal Q-function is simply the maximum Q-function over all policies and also obeys the Bellman Equation, as $Q^*(s,a) = \max_{a \in \mathcal{A}} Q^{\pi}(s,a) = r(s,a) + \gamma \max_{a \in \mathcal{A}} Q^*(s', a')$. The state-action pair (s', a') refers to any possible next state-action pair. To find the optimal policy, $Q^*(s,a)$ must be found. A solution to this problem is given by the Q-learning algorithm [12].

2.2 Q-Learning and Deep Q-Learning

In Q-learning, the Q-function is represented using a lookup table that stores the expected rewards for each state-action pair. The table entries are updated with each action that the agent takes, as $Q(s_t, a_t) \leftarrow r(s_t, a_t) + \gamma \max_{a_{t+1}} Q(s_{t+1}, a_{t+1})$. Actions are chosen by observing the state and the corresponding Q-value estimates in the Q-table, with the action that maximizes the Q-values being chosen.

To ensure adequate learning in the environment, agents must balance exploring the environment and exploiting their policy. This is done via the $\epsilon-$greedy strategy, where the agent will explore with a random action $a' \in \mathcal{A}$ with probability ϵ or exploit past learning using $a = \text{argmax}_{a \in \mathcal{A}} Q(s, a')$ with probability $1 - \epsilon$.

Q-learning is impractical in real-life situations, as the state space $\mathcal{S}$ can be huge, resulting in a prohibitively large Q-table. This issue of scalability can be solved by Q-function approximation with a deep neural network.

A deep neural network that approximates the Q-function is called a *deep Q-Network*, or DQN. The state s_t is input to the network and the output size is equal to the size of the action space $\mathcal{A}$. The action that corresponds with the largest value is selected to be executed. This neural network can also be called the *policy-net* θ_p, as it represents the optimal policy.

To move the approximation of the DQN closer to the optimal Q-function, the MSE loss is found at time t using a second DQN, called the *target-net* θ_t. Both the policy and target-net Q-values are calculated using the Bellman Equation, as $Q(s_t, a_t, \theta) = r_t + \gamma \max_{a_{t+1}} Q(s_{t+1}, a_{t+1}, \theta)$. The loss is back-propagated and the internal weights of the policy-net updated. After a sufficient amount of updates, the policy-net has moved in the direction that minimises the loss, but the target-net has not. After a number of time steps, the weights of the policy-net are copied to the target-net to update Q^{target} with respect to what has been learnt already.

3   QoS-Aware Reinforcement Learning Scheduler for LTE Networks

The static nature of the scheduler's policy limits the efficacy of the scheduler to cater for the dynamics of the network, in terms of mixed traffic types, UE numbers and CQI. Enabling the smart swapping of the scheduler's policy would support increased performance, by considering other network indicators.

3.1   QoS Requirements

To facilitate the DQN possibly changing the packet scheduler every TTI, some changes are introduced to how QoS are managed by the radio bearers and application flows.

Application flows define a set QoS requirements for dedicated-bearer registration at the eNodeB. The application is assigned one of the QoS parameter classes based on the selected scheduler type. LTE-Sim offers a single *QoSParameters* base object and three additional derived classes *QoSForM_LWDF*, *QoSForFLS* and *QoSForEXP*. The derived classes provide additional functionality for M-LWDF, FLS and EXP packet schedulers respectively, while other scheduler types utilise the base class [10].

The base object *Application* is modified to accept all four *QoSParameter* objects so that functionality may continue when the packet scheduler is changed. An additional helper function is implemented, *Application::UpdateQoS()*, to ensure that the correct *QoSParameter* object is selected. The function also updates the radio bearer that is associated with the application flow to the new *QoSParameter* object.

3.2   Simulation Output Trace

The trace of the simulation is output directly to the standard output. At the end of each TTI, transmit and receive events are output to the trace. It is possible that there are no events for that TTI, and this case is referred to as an invalid update. To capture the output and process it, the standard output stream is redirected to a *std::stringstream*.

Using the output trace, the measured packet-loss rate, bit-rate and delay can be calculated. The PLR for flow j at TTI t is calculated as $PLR_j(t) = 1 - \frac{rec_j(t)}{sent_j(t)}$, where $rec_j(t)$ and $sent_j(t)$ are the number of all sent and received packets to flow j until TTI t. If all packets that have been sent are received, then the PLR will be 100%.

For the bit-rate calculation, a FIFO buffer of 1000 TTIs is maintained, where the elements hold the RX messages from TTI $t - 1000$ to the current TTI t. To calculate the bitrate for flow j, the buffer is iterated over and the packets that were received by flow j have their sizes accumulated, as $buffer_j$. Then, the bitrate at TTI t is found as $bitrate_j(t) = \frac{8*buffer_j}{1000}$. The accumulated packet sizes are stored in bytes and therefore converted to bits with a multiplication.

Since each TTI measures 1 ms, 1000 TTIs constitutes 1 s. The final bit-rate is given in *bits/second*. To measure the experienced delay of the flow, RX events for flow j are used directly as $Delay_j(t)$.

4 Deep Q-Network Implementation

At each TTI, the goal of the agent is to pick and use the most suitable packet scheduler that maximizes the number of satisfied users regarding their QoS requirements. A generalized mathematical problem statement of this coarse-grained maximization objective is covered in [13].

4.1 State, Action, and Rewards

To understand the performance of the network, the indicator vector, $I_j(t)$, for flow j at TTI t is defined as: $I_j(t) = [bitrate_j(t), Delay_j(t), PLR_j(t)]$. The components of the vector are the measured indicators, respectively. The model of a 4G-LTE network is complex, involves many dynamic elements and is not trivially generalized, thus this research chooses to use a model-free reinforcement learning approach. Furthermore, we use a deep Q-network to approximate the Q-function using a neural network. The output of the DQN agent is equal to the size of the action space, where the values represent the expected Q-values for each action. The largest Q-value is selected, and the corresponding scheduler is selected for the next TTI.

Furthermore, the overall reward at TTI t, having taken action a_t from state s_t, and transitioning to state s_{t+1} is found using $r_t = \frac{1}{N}\sum_{j=0}^{N}(\alpha\,R_{j0} + \beta\,R_{j1} + \theta\,R_{j2})$. Here, N is again the number of application flows to be scheduled. Using co-efficients for each sub-reward, the relative importance of each requirement can be dictated. The sub-reward for each flow, R_j, is calculated with a component for each QoS requirement, x_{j0}, x_{j1}, x_{j2}:

$$R_{j0} = \begin{cases} \dfrac{bitrate_j(t)}{x_{j0}}, & \text{if } bitrate_j(t) \leq x_{j0} \\ 1, & \text{otherwise} \end{cases} \qquad R_{j1} = \begin{cases} 1, & \text{if } Delay_j(t) \geq x_{j1} \\ 0, & \text{if } Delay_j(t) \leq x_{j1} \\ -0.1, & \text{if } flow_j \text{ has not received any packets} \end{cases}$$

$$R_{j2} = \begin{cases} 1, & \text{if } PLR_j(t) \geq x_{j2} \\ 0, & \text{if } PLR_j(t) \leq x_{j2} \\ -0.1, & \text{if } flow_j \text{ has not received any packets} \end{cases}$$

Considering that the bitrate is calculated using a buffer method, a linear reward function is used to encourage the agent to move towards the requirement. The sparse rewards for PLR and delay are used as they have shown to be easier to achieve in prior simulations.

Using the scheduler, the action space of the agent is only 6 actions. To decide on which action to make (or what scheduler to pick for the next TTI) the state at TTI t is defined as the following vector: $s_t = [N,\ I_0(t),\ I_1(t),\ldots,\ I_N(t),\ CQI_0,\ CQI_1,\ldots,\ CQI_N]$. N is the number of application flows in the network and CQI_j is the reported CQI values for flow j. The CQI values are estimations for the signal-to-noise ratio for each downlink subchannel.

4.2   Training Method and Simulation Architectures

To train the DQN agent, the method used in shown in Algorithm 1. All calculations involving the policy and target neural networks are done with the GPU. The DQN is trained in epochs until the soft-divergence condition is met [14], i.e. once $max\,(policy\_net\,(s_t)) \geq 100$. The replay memory is only updated when a valid update is received from LTE-Sim. Additionally, the DQN is only trained every four valid updates to allow the environment to develop [15].

Figure 1 shows the configuration between LTE-Sim and Pytorch. The following message passing functions for LTE-Sim are implemented:

- *SendUESummary()*: Open pipe A and send each application ID, their associated UE and the QoS requirements. Only called on initialisation.
- *SendState()*: Open pipe A and send the redirected output trace.
- *SendCQISummary()*: Open pipe B and send the CQI of all UES.
- *GetScheduler()*: Open pipe C and receive the scheduler that was passed by the DQN. Call *Application::UpdateQoS()* for all flows.

Reciprocating message passing functions must also be implemented for the Pytorch-DQN framework. All message passing functions block while waiting for the payload to be delivered.

Algorithm 1: DQN agent training

---

initialise *replay_memory, policy_net, target_net, state, update_time*
initialise γ, ϵ, lr, bs, $x = 0$
while *not_diverged* **do**
  increment x and get the current state s_t
  select a scheduler $a_t = \mathrm{argmax}_a$ policy_net(s_t, ϵ)
  give scheduler to LTE-Sim
  observe the next state s_{t+1} and calculate reward $r_t\,(s_{t+1})$
  store experience $e_t = (s_t, a_t, r_t\,(s_{t+1}), s_{t+1})$ in replay_memory
  get random batch $B = [s_b, a_b, r_b, s_{b+1}]$ of size bs from replay_memory
  calculate $Q(s_b, a_b) = $ policy_net(s_b)
  calculate $Q^{target}(s_b, a_b) = r_b + \gamma$ target_net(s_{b+1})
  find MSE loss between $Q(s_b, a_b)$ and $Q^{target}(s_b, a_b)$
  perform gradient descent on policy_net to minimise loss
  if $x == update\_time$ **then**
    copy weights of policy_net to target_net
    x = 0
  end
  if $max\,(Q\,(s_b, a_b)) \geq 100$ **then**
    not_diverged = false
  end
end

---

5   Evaluation and Results

5.1   Methodology and System Parameters

To evaluate the proposed system as discussed in Sect. 3 an Ubuntu 18.04.3 LTS system with 256 GB of memory, an Intel Xeon Gold 6132 (2.6 GHz) and a NVIDIA GeForce RTX 2080 Ti is used. The simulation parameters for LTE-Sim and the DQN are shown in Table 1. The DQN is trained in epochs consisting of 25,000 TTIs until the soft-divergence condition is met. The replay memory has a capacity of 30,000 experiences and ensures complete state-action exposure. The reward coefficients are set $\alpha = 0.6$ and $\beta = \theta = 0.2$. A large α value is chosen to further encourage the promotion of bitrate.

Three scenarios with differing traffic types, while increasing the numbers of UEs, are explored. At each TTI, the PLR, delay and bitrate of each flow are measured using the equations in Sect. 3.2. These measured indicators are compared to the respective QoS requirements and marked as either satisfied or unsatisfied in that TTI. Using the count of satisfied TTIs for each requirement, the average satisfaction rates over all UEs of the same traffic type is recorded. In the mixed traffic scenario C, the average is taken for both traffic types separately. Additionally, the cumulative size of all the received packets for each UE in the scenario is recorded, which is referred to as the goodput.

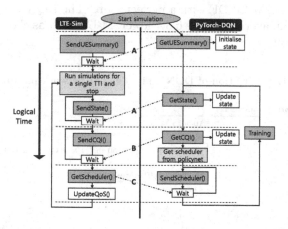

Fig. 1. Communication flow diagram between LTE-Sim and Pytorch. Each send function waits for its reciprocal receive function before continuing.

5.2   Experimental Results

The results for scenarios A and B are shown in Tables 2 and 3 respectively. In all scenarios, the GBR requirement was never met and therefore the goodput is used as a measure of bitrate. All satisfaction rates and goodputs are normalised vertically to the highest performing scheduler in the QoS category. The rows in grey are the schedulers that the DQN chose in the scenario for that number of UEs. The average score of each scheduler is shown in the last column for that number of UEs, with the largest average being the best overall performer. The satisfaction rates and goodputs are not normalised, to highlight the differences between the traffic types.

Scenario A. With 50 UEs, the top three performing schedulers were EXP, the DQN and M-LWDF. The EXP scheduler was the top performer for both the PLR and goodput requirements. The DQN was a close second, with only a 2.2% difference in the overall score. The DQN chose the EXP scheduler for the majority of the experiment. With 100 UEs, the DQN was the top performer for the delay and goodput requirements, with the PF scheduler being the best performer for the PLR requirement. The DQN was the best performer all-round, with a 5.7% improvement over the next-best scheduler, EXP. For 100 UEs, the DQN chose the M-LWDf scheduler. At 400 UEs, the best overall performer was Opt. EXP, with the DQN coming second with a 0.9% difference in average score. The DQN was the best performer in PLR and picked the FLS scheduler.

Scenario B. In the scenario with 50 UEs, the best overall performer was M-LWDF, which performed best in delay and goodput. The DQN only achieved an average score of 0.775, placing it as the worst performer. With 100 UEs, the best performer was Opt. LOG, with a perfect score. The DQN has an average score that is 12.2% lower, and selected the EXP scheduler. With 400 UEs, the best performer was M-LWDF with a perfect score. The DQN was the third best performer, with an average that is 6.3% lower. M-LWDF was chosen by the DQN for the experiment.

Table 1. Description of DQN and LTE-Sim network parameters.

LTE-sim network parameters	
Bandwidth	10 Mhz
Cell Radius	1000 m
QoS Requirements	Video - 180 kbps, 150 ms, 10e-3 CBR - 256 kbps, 300 ms, 10e-6
CBR flows	Interval - 0.01 s Message Size- 320 B
Video Flows	440 k trace
DQN Parameters	
Reward co-efficients α, β, θ	0.6, 0.2, 0.2
Discount Factor γ	0.999
Batch Size	32
Replay Memory Size	30000
Epsilon	1.0–0.01 (decay rate of 0.00005)
Training Duration	25000 TTIs
Optimiser	Adam
Learning Rate	0.001
DQN layers	Input: linear, 80 nodes hidden: hyperbolic tangent output: linear, 80→ 6
Learning Frequency	4 valid updates
Target Network Update Interval	40 valid updates

Scenario C. With 50 UEs, FLS performs well in terms of the video flows's delay satisfaction and goodput. However, it does not offer the same performance for the CBR flows. Overall, the video flow's satisfaction rates are low. PF,EXP and Opt. LOG perform well with the CBR flows. The DQN performs moderately, favouring the Opt. LOG scheduler.

Table 2. Scenario A results.

	50				100				400			
	PLR	Delay	Goodput	Ave.	PLR	Delay	Goodput	Ave.	PLR	Delay	Goodput	Ave.
PF	0.918	0.449	0.499	0.622	1	0.331	0.437	0.589	0.968	0.086	0.324	0.459
M-LWDF	0.934	0.955	0.954	0.948	0.513	0.942	0.91	0.788	0.745	0.849	0.822	0.805
EXP	1	0.995	1	0.998	0.548	0.992	0.978	0.839	0.775	0.78	0.744	0.766
FLS	0.985	0.868	0.867	0.907	0.456	0.835	0.771	0.687	0.756	0.914	0.913	0.861
Opt. EXP	0.136	1	0.987	0.708	0.326	0.947	0.915	0.729	0.805	1	1	0.935
Opt. LOG	0.59	0.999	0.981	0.857	0.573	0.981	0.957	0.837	0.637	0.776	0.743	0.719
DQN	0.985	0.956	0.959	0.967	0.688	1	1	0.896	1	0.891	0.888	0.926

With 100 UES, the DQN provides the best performance for the CBR flow's delay and goodput, but not PLR. The best PLR is supplied by PF fair. The video flows's satisfaction rates are again low, with EXP giving the best delay and PLT. In terms of video goodput, the best performance is given again by EXP. The DQN provides moderate performance for both flow types, favouring the EXP scheduler.

With 400, the video flow's satisfaction rates are very low and the goodput is lower than that of the CBR flows. The DQN provides the best performance in terms of CBR delay and goodput, as well as the average goodput across both traffic types. EXP shows the best CBR performance.

Table 3. Scenario B results.

	50				100				400			
	PLR	Delay	Goodput	Ave.	PLR	Delay	Goodput	Ave.	PLR	Delay	Goodput	Ave.
PF	1	0.745	0.747	0.831	0.822	0.613	0.698	0.771	0.981	0.144	0.648	0.591
M-LWDF	0.868	1	1	0.956	0.754	0.817	0.818	0.796	1	1	1	1
EXP	0.713	0.83	0.838	0.794	0.864	0.894	0.889	0.882	1	0.933	0.938	0.957
FLS	0.856	0.963	0.979	0.933	0.828	0.982	0.957	0.922	0.892	0.948	0.951	0.93
Opt. EXP	0.767	0.95	0.936	0.884	0.662	0.901	0.89	0.818	0.667	0.847	0.851	0.788
Opt. LOG	0.656	0.888	0.91	0.818	1	1	1	1	0.888	0.901	0.905	0.898
DQN	0.719	0.788	0.819	0.775	0.803	0.912	0.92	0.878	0.907	0.971	0.98	0.953

Inference Complexity. The inference times of the policy network for all experiments, with 400 UEs is shown in Fig. 3. Across all scenarios, the inference time at 400 UEs is less than 0.25 ms. This is well below the timing requirements of 1 ms. The DQN implementation is both scalable and easily achieved (Figs. 2 and 3).

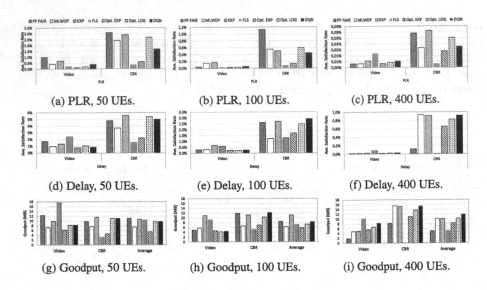

Fig. 2. Scenario C results.

5.3 Discussion of Results

The results from scenarios A show that the DQN provides good performance for video flows. At all number's of UEs, the DQN favoured the top overall performers. At 50 UEs, the delay and PLR requirement was mostly satisfied. However, as the number of UEs increased to 100 and 400, the delay requirement became harder satisfy.

Scenarios B's results show that the DQN could only provide moderate overall performance for CBR flows at 50 and 400 UEs. However, at 400 UEs, the DQN provides good performance in delay and PLR satisfaction, as well as goodput. The DQN is able to select the best scheduler at 400 UEs, but favours the moderately performing schedulers at 50 and 100 UES. CBR's delay requirement is easier to satisfy than the video's requirement. While the PLR requirement is much harder to satisfy as the number of UE's increases.

In scenario C, the PLR and delay satisfaction rates of the CBR flows are overall higher than the video flows. This means that they can provide more

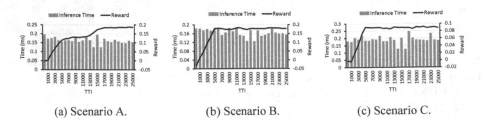

Fig. 3. Inference time and reward in all scenarios with 400 UEs.

reward for the DQN to pursue. At 50 UEs the DQN favoured the Opt. LOG scheduler for it's high performance for CBR flows and moderate video performance. At 100 UEs the DQN favoured the EXP scheduler, that provided the best performance of video PLR, delay and goodput, as well as CBR delay and goodput. Lastly, at 400 UEs, the EXP scheduler was again favoured by the DQN for high performance in CBR's PLR and delay satisfaction, and goodput. The DQN captured the dynamics of the mixed-traffic network well and was able to optimise the QoS satisfaction and goodput of the mixed flows.

It is also clear that the DQN's performance is bounded by the performance of the schedulers that occupy the action space. In regard to QoS requirement satisfaction, the DQN cannot improve performance to levels above the scheduler's ability. The DQN training could incur additional delay to the overall system. However, we find the inference delay falls under 1 ms which is within a single TTI once the training is completed. In comparison with the existing schedulers, the DQN was able to handle the dynamics of the mixed-traffic network and maximize the level of satisfaction across both traffic types using the available schedulers.

6   Related Work

High-level Methods. This group uses a reinforcement learning agent to pick a typical scheduling policy every allocation interval. This research falls into this category. In [13], the actor-critic (A3C) reinforcement learning algorithm is used to train an agent that picks a scheduler, to maximize QoS satisfaction. The action space is composed of three PF-derived schedulers. The evaluation of their method does not include mixed traffic types or a high user density. Similarly, [16] uses the deep deterministic policy gradient method to make the same decisions in a very large action space.

Low-level Methods. This category uses a reinforcement learning agent to fill the RB allocation map. These methods have a action space composed of all the active flows in the network and make fine grained allocation decisions for each RB. In [17], the authors use the PF scheduler to guide a DDPG agent in fair RB allocation decisions. This solution does not allow the agent to develop its own policy. The DDPG agent in [3] tries to optimise fairness and throughput. Both of these works only evaluate their systems using a small number of UEs and RBs, not demonstrating scalability. Additionally, an alternative method that directly approximates the policy of the agent using a deep neural network is given in [18].

7   Conclusion

This paper presents a reinforcement learning solution to manage radio resource allocation in 5G and future mobile networks. By using the Q-learning method, this research approximates the Q-function with a deep neural network. The DQN agent can produce a sequence of coarse-grained scheduling decisions that maximizes the satisfaction rates of PLR, delay and bitrate QoS requirements. In scenarios of single- and mixed-flow network traffic, the DQN agent can select the scheduler that provides the best overall performance for all flows. At each TTI, the optimal scheduler is selected, with minimal computational overhead and complexity. The reward model prioritizes bitrate, but provides flexibility to promote any of the requirements. Compared to other research, this paper's approach uses a simple deep Q-network RL algorithm which shows stable training and inference performance. Additionally, this paper shows that the proposed method is both light-weighted computation and scalable.

Acknowledgements. This research was in part supported by the MOTIE (Ministry of Trade, Industry & Energy) (No. 10080674, Development of Reconfigurable Artificial Neural Network Accelerator and Instruction Set Architecture) and KSRC (Korea Semiconductor Research Consortium) support program for the development of the future semiconductor device and in part supported by Samsung Electronics Co., Ltd.

References

1. "Cisco: 2020 CISO Benchmark Report," Computer Fraud & Security, vol. 2020, no. 3, p. 4 (2020)
2. Barnett, T.J., Sumits, A., Jain, S., Andra, U.: Cisco Visual Networking Index (VNI) update global mobile data traffic forecast. Vni, pp. 2015–2020 (2015). http://www.cisco.com/c/en/us/solutions/collateral/service-provider/visual-net working-index-vni/complete-white-paper-c11-481360.html
3. Al-Tam, F., Correia, N., Rodriguez, J.: Learn to Schedule (LEASCH): a deep reinforcement learning approach for radio resource scheduling in the 5G MAC layer. IEEE Access **8**, 108-088–108-101 (2020)
4. Shafi, M., et al.: 5G: a tutorial overview of standards, trials, challenges, deployment, and practice. IEEE J. Sel. Areas Commun. **35**(6), 1201–1221 (2017)
5. Sallab, A.E., Abdou, M., Perot, E., Yogamani, S.: Deep reinforcement learning framework for autonomous driving. arXiv, pp. 70–76 (2017)
6. Levine, S., Pastor, P., Krizhevsky, A., Ibarz, J., Quillen, D.: Learning hand-eye coordination for robotic grasping with deep learning and large-scale data collection. International Journal of Robotics Research **37**(4–5), 421–436 (2018)
7. Mnih, V., et al.: Human-level control through deep reinforcement learning. Nature **518**(7540), 529–533 (2015). http://dx.doi.org/10.1038/nature14236
8. Vinyals, O., et al.: Grandmaster level in StarCraft II using multi-agent reinforcement learning. Nature **575**(7782), 350–354 (2019). http://dx.doi.org/10.1038/s41586-019-1724-z
9. Paszke, A., et al.: PyTorch: an imperative style, high-performance deep learning library, arXiv (2019)
10. Piro, G., Grieco, L.A., Boggia, G., Capozzi, F., Camarda, P.: Simulating LTE cellular systems: an open-source framework. IEEE Trans. Veh. Technol. **60**(2), 498–513 (2011)
11. Szepesvári, C.: Algorithms for reinforcement learning. Morgan & Claypool Publishers, vol. 9 (2010)
12. Watkins, C.J., Dayan, P.: Q-learning. Mach. Learn. **8**(3–4), 279–292 (1992)
13. Comsa, I.S., De-Domenico, A., Ktenas, D.: QoS-driven scheduling in 5g radio access networks - a reinforcement learning approach. In: 2017 IEEE Global Communications Conference, GLOBECOM 2017 - Proceedings, vol. 2018-Janua, pp. 1–7 (2017)

14. van Hasselt, H., Doron, Y., Strub, F., Hessel, M., Sonnerat, N., Modayil, J.: Deep Reinforcement Learning and the Deadly Triad. arXiv (2018)
15. Fedus, W., et al.: Dabney. Revisiting Fundamentals of Experience Replay, arXiv (2020)
16. Tseng, S.C.. Liu, Z.W., Chou, Y.C., Huang, C.W.: Radio resource scheduling for 5G NR via deep deterministic policy gradient. In: 2019 IEEE International Conference on Communications Workshops, ICC Workshops 2019 - Proceedings (2019)
17. Wang, J., Xu, C., Huangfu, Y., Li, R., Ge, Y., Wang, J.: Deep reinforcement learning for scheduling in cellular networks. arXiv (2019)
18. Shekhawat, J.S., Agrawal, R., Shenoy, K.G., Shashidhara, R.: A reinforcement learning framework for QoS-driven radio resource scheduler. In: 2020 IEEE Global Communications Conference, GLOBECOM 2020 - Proceedings, pp. 1–7 (2020)

Adaptive Buffering Scheme for PCM/DRAM-Based Hybrid Memory Architecture

Xiaoliang Wang[1], Kaimeng Chen[2(✉)], and Peiquan Jin[1]

[1] University of Science and Technology of China, Hefei 230027, China
wxll47@mail.ustc.edu.cn, jpq@ustc.edu.cn
[2] JiMei University, Xia'Men 361021, China

Abstract. Phase Change Memory (PCM) motivates new hybrid memory architecture that consists of PCM and DRAM. A critical issue in the PCM/DRAM-based hybrid memory architecture is how to devise efficient buffering schemes for heterogeneous memory. A feasible way is to use PCM as a page buffer in the memory hierarchy. Based on this assumption, this paper presents a new buffering scheme named FWQ-LRU for PCM/DRAM-based hybrid memory. FWQ-LRU proposes two new designs. First, it uses two queues for DRAM to identify write-cold DRAM pages, which will be moved to PCM during page replacement and page migrations. Second, to reduce PCM writes, it adopts an adaptive page migration mechanism to swap write-hot PCM pages with write-cold DRAM pages. Specially, FWQ-LRU devises a queue called FWQ (FIFO Write Queue) to detect write-hot PCM pages. Further, we periodically measure the page-migration benefits under the current workload to adjust the page migration policy for future requests. With this mechanism, write requests are mainly redirected to DRAM, and the writes to PCM are reduced. In addition, FWQ-LRU can maintain the same hit ratio as LRU. We conduct comparative trace-driven experiments and the results suggest the efficiency of our proposal.

Keywords: Buffer management · PCM · Hybrid memory · LRU

1 Introduction

The increasing needs for large main memories call for new types of memories, such as flash memory and Phase Change Memory (PCM). While flash memory only supports page-based I/Os, PCM is byte-addressable and can be used as an alternative to DRAM. In addition, compared with DRAM, PCM is non-volatile and has higher storage density and lower energy consumption. Thus, many researchers have proposed to incorporate PCM into the memory hierarchy of computer systems [1–4].

However, two problems of PCM make it challenging to replace DRAM in current computer systems. First, the write latency of PCM is about 6 to 50 times slower than that of DRAM [5, 6], as shown in Table 1. Second, PCM has a worn-out problem, i.e., each PCM cell has limited write endurance. Thus, PCM is not suitable for write-intensive applications. Therefore, a more practical way to utilize PCM in the memory architecture is to use both PCM and DRAM to construct hybrid memory architecture

© IFIP International Federation for Information Processing 2022
Published by Springer Nature Switzerland AG 2022
C. Cérin et al. (Eds.): NPC 2021, LNCS 13152, pp. 118–130, 2022.
https://doi.org/10.1007/978-3-030-93571-9_10

[4, 7, 8], as shown in Fig. 1. This hybrid architecture organizes PCM and DRAM at the same level in the main memory. However, since PCM's read/write latency is still lower than that of DRAM, PCM is not suitable to replace DRAM as process memory space. In other words, it is more practical to be used as a page buffer for files or databases. This yields a hybrid page buffer including PCM and DRAM.

This paper focuses on the buffer management for data pages cached by the hybrid memory consisting of PCM and DRAM. In particular, we concentrate on page replacement algorithms for the hybrid memory shown in the right part of Fig. 1. More specifically, our work aims to improve the performance of the hybrid buffer over I/O-intensive workloads. Unlike traditional page replacement algorithms designed for DRAM-only main memory [9], such as LRU and CLOCK, page replacement algorithms for the hybrid memory architecture have to be aware of the different characteristics of PCM and DRAM. To be more specific, page replacement algorithms for hybrid memory need to consider hit ratio and the write operations to PCM.

Table 1. PCM vs. DRAM [6, 7]

	DRAM	PCM
Durability	Volatile	Non-volatile
Read Latency (per *cacheline*)	20~50 *ns*	20~100 *ns*
Write latency (per *cacheline*)	20~50 *ns*	300~1000 *ns*
Density	Low	High (~ 4X DRAM)
Endurance	∞	~ 10^{10} for writes

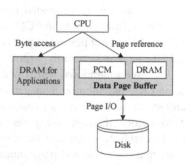

Fig. 1. Hybrid memory architecture consisting of PCM and DRAM

This paper presents an efficient approach called FWQ-LRU (FIFO-Write-Queue LRU) for buffer management over the PCM/DRAM-based hybrid memory architecture. FWQ-LRU aims to keep a high hit ratio and, at the same time reducing PCM writes. Briefly, we make the following contributions in this paper:

(1) We propose a new buffer management algorithm, FWQ-LRU, for the hybrid memory architecture based on DRAM and PCM. FWQ-LRU can identify write-intensive pages and put them in DRAM to reduce PCM writes. FWQ-LRU adopts an adaptive page migration mechanism to adapt to workload changes. With this mechanism, it can maintain a high reduction rate of PCM writes under different workloads.

(2) We perform trace-driven experiments to evaluate the performance of our proposal. We use different types of workloads and make comparisons with several previous LRU-based methods. The results show that our proposal outperforms its competitors in reducing PCM writes.

2 Related Work

Seok et al. [10] proposed a page replacement algorithm named Hybrid-LRU that integrates LRU with a prediction technique and a page-migration mechanism. Hybrid-LRU sets a weight value for each page in the memory. Each time a page is hit, its weight value is re-calculated according to the reference type. Hybrid-LRU uses two thresholds to judge whether the referenced page becomes read-intensive or write-intensive. Read-intensive pages in DRAM are moved to PCM, and write-intensive pages in PCM are moved to DRAM. However, for read-intensive workloads, Hybrid-LRU will incur many PCM writes caused by the page migration operations which move pages from DRAM to PCM. MHR-LRU [11] is another LRU-based buffer management scheme for the DRAM/PCM-based hybrid memory architecture. MHR-LRU always places the most recently written pages in DRAM to reduce PCM writes. When a page fault occurs, MHR-LRU selects the least recently used page as the victim. If the victim is a PCM frame, but the requested page is referenced by a write request, instead of placing the requested page on PCM, MHR-LRU moves the least recently written page from DRAM to PCM and places the requested page in DRAM. However, MHR-LRU will not perform a page migration unless a page replacement occurs. Thus, if a PCM page is frequently written and becomes write-intensive, MHR-LRU will not perform page migrations, which will result in many writes to PCM. APP-LRU [12] is also an LRU-based buffer management algorithm that was proposed for hybrid memory architecture. It introduces a metadata table to record the access history of pages in the heterogeneous memory composed of PCM and DRAM. APP-LRU first uses the metadata table to determine the memory place (PCM or DRAM) when a page fault occurs that the requested page should be placed in. Then, it selects the least recently used page as the victim. If the victim does not reside in the same memory place as the requested page, APP-LRU releases the victim and performs a page migration between PCM and DRAM to get a free frame for the requested page. APP-LRU will introduce many PCM writes under read-intensive workloads because it always keeps read-intensive pages in PCM. In addition, APP-LRU only performs page migrations when a page replacement occurs; thus, it has the same problem as MHR-LRU.

D-CLOCK [13] is a CLOCK-based buffer management scheme for hybrid memory architecture. Like MHR-LRU and APP-LRU, D-CLOCK also performs a page migration only when a page fault occurs. D-CLOCK modifies the CLOCK algorithm when selecting the victim. If a page in PCM is selected as the victim many times, D-CLOCK will only select a page from DRAM as the victim for the next page fault. This will lower the hit ratio compared with the traditional CLOCK algorithm. CLOCK-RWRF [14] and CLOCK-DWF [15] are also CLOCK algorithms improved for PCM and DRAM-based hybrid memory.

3 Design of FWQ-LRU

3.1 Overview

FWQ-LRU is an LRU-based algorithm aiming to reduce write operations to PCM but without degrading the hit ratio. Figure 2 shows the structure of FWQ-LRU. FWQ-LRU employs an LRU list to manage all the pages (including DRAM pages and PCM pages) in the hybrid memory. This LRU list is the same as that in the traditional LRU algorithm. In addition, DRAM pages are also maintained in two write queues, namely a hot queue and a cold queue, which are used for page migrations, but each DRAM page can only be in one queue. The hot queue and the cold queue are designed to record write-request information of each page in DRAM. This kind of information is further used to detect write-cold pages in DRAM. The detected write-cold pages will be moved to PCM by the page migration algorithm.

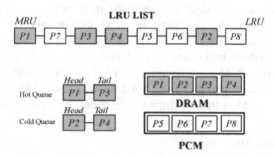

Fig. 2. Structure of FWQ-LRU

Algorithm 1. FixPage (r)
Input: a page request R
Output: a free frame in the hybrid memory
1:   **if** there is a free frame s in DRAM **then**
2:         **return** s;
3:   **else if** there is a free frame s in PCM **then**
4:         **if** r is a read request **then**
5:               **return** s;
6:         **else**
7:               $m =$ the page in the tail of cold queue
8:               migrate m to s;
9:               **return** a free frame in DRAM;
10: **else**
11:       get $victim$ from LRU position of the LRU list;
12:       **if** $victim$ is in PCM and q is a write **then**
13:             $m =$ the page in the tail of cold queue;
14:             release $victim$ and migrate m to the free frame in PCM;
15:             **return** a free frame in DRAM;
16:       **else**
17:             release $victim$;
18:             **return** a free frame in PCM;

FWQ-LRU sets a hotness bit (*hot_bit*) for each page in DRAM. When a page fault occurs (meaning that the requested page is neither in DRAM nor in PCM) and the requested page is in DRAM, we perform different operations according to the request type. If the request is a write request, the page is inserted into the head of the cold queue, and its *hot_bit* is set to 1; if the request is a read request, the page is inserted into the tail of the cold queue and its *hot_bit* is set to 0. When a write request hits a DRAM page in the cold queue, if its *hot_bit* is 0, we move the page to the head of the cold queue and set its *hot_bit* to 1; if the page's *hot_bit* is 1, the page is moved to the head of the hot queue. The pages close to the tail of the cold queue are considered as write-cold pages, which will be moved to PCM in case of page faults. In other words, pages in the hot queue are regarded as write-hot pages, and we expect to maintain them in DRAM.

When a page fault occurs, FWQ-LRU first needs to determine where the requested page should be placed. If there is a free frame in DRAM, FWQ-LRU simply returns the free frame. However, if there is a free frame in PCM, we need to consider the request type. When the request is a read request, the free frame in PCM is returned. If the request is a write request, FWQ-LRU selects the page in the tail of the cold DRAM queue, moves the page to the free frame in PCM, and returns the selected frame in DRAM as the free frame. If there is no free frame in the hybrid memory, the least recently used page in the hybrid memory is selected as the victim, which is the same as LRU. If the victim is in PCM and the request is a write request, FWQ-LRU releases the victim, moves the page in the tail of the hot DRAM queue to PCM, and returns the selected frame in DRAM as the free frame. The details are shown in Algorithm 1.

3.2   Adaptive Page Migration

FWQ-LRU performs page migrations when a page in PCM is hit by a write request. The basic rule of page migration is to move future write-intensive pages to DRAM and move future read-intensive pages to PCM. In this paper, we adopt a *write-interval-based second-chance* policy to detect the write intensity of PCM pages. The write interval of a PCM page refers to the interval between two adjacent writes to the page. The reason of using second chance but not third or fourth chance is based on the observation in [16]. As a consequence, our policy considers both the write temporal locality and the write interval to predict the write intensity of PCM pages. The write-interval-based second-chance policy works as follows. FWQ-LRU maintains a parameter "*migration interval*"; when a write request hits a page in PCM, and it checks the interval between this write request and the last write request to the page. If the interval is less than the *migration interval*, we regard the page as a write-hot page.

To implement the write-interval-based second chance policy, we introduce a data structure called *FIFO Write Queue* (FWQ) to record the write-access information to pages. Figure 3 shows the structure of FWQ. When a write request comes to the hybrid memory, the requested page number is pushed into the head of FWQ. If FWQ is full, the record in the tail of FWQ will be removed. Consequently, FWQ contains the most recent n write requests. When a write request hits a page in PCM, FWQ-LRU checks if the requested page has a record in FWQ. If the page does not have a record in FWQ, we know that the interval between this write request and the last write request to the page is larger than the size of FWQ. Conversely, if the requested page has a record in FWQ, we know that the interval between this write request and the last write request to the page is less than the size of FWQ. Thus, we can adjust the migration interval by changing the size of FWQ.

3.3 Adjusting the FWQ Size

The FWQ size affects the efficiency of page migrations, and an optimal FWQ size for current workloads is needed to be determined. To solve this problem, we periodically measure the current effect of page migration and adjust the FWQ size to optimize the impact of future page migration.

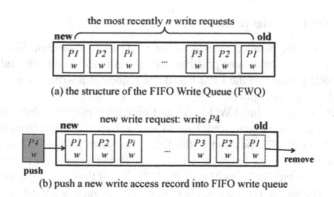

Fig. 3. The structure and management of FWQ

To determine whether to perform a page migration, we first compute the effect of page migrations. For all pages that are initially placed in PCM, we define the total number of write requests that hit these pages in DRAM as the *"positive effect"* of page migrations. For all pages that are initially placed in DRAM, we define the total number of write requests that hit these pages in PCM as the *"negative effect"* of page migrations. A positive effect is helpful for reducing PCM writes, while a negative effect will increase the writes to PCM. Then, we can define the reduced PCM writes as the difference between positive effect and negative effect. We set a *memory_bit* for each page to mark the initial memory location of the page. If the page is placed on DRAM, *memory_bit* is set to 1, otherwise, *memory_bit* is set to 0. Therefore, we define *positive effect* as the number of the write requests hitting in DRAM pages with *memory_bit* = 0 and *negative effect* as the number of the write requests hitting in PCM pages with *memory_bit* = 1.

Next, we consider the additional cost of page migrations. FWQ-LRU performs page migrations only when a write request hits a page in PCM. If the page hit by a write request is needed to be moved to DRAM, FWQ-LRU exchanges this page with a DRAM page in the cold DRAM queue. This page exchange introduces one PCM write operation, and one DRAM write operation. After that, the write request is redirected to DRAM, which incurs another DRAM write operation. Therefore, the additional cost of a page-migration operation is two DRAM write operations. Then, we can define the extra cost by the increased DRAM writes.

After that, we can define the benefits of page migrations as *PCM_Write_Reduce* * *PCM_Write_Cost* − *DRAM_Write_Increase* * *DRAM_Write_Cost*. Here, *PCM_Write_Cost* is the cost of a write operation to PCM, and *DRAM_Write_Cost* is the cost of a write to DRAM. Both of them can be computed by write latency. FWQ-LRU adopts a cost-based adaptive adjustment policy to adjust the FWQ size by periodically measuring page migration benefits.

3.4   The FWQ-LRU Algorithm

Algorithm 2 shows the detailed process of the FWQ-LRU algorithm. When a DRAM page is hit, FWQ-LRU calculates positive effects and negative effects, and moves the page to the MRU position of the LRU list. If the request is a write request, its position in the DRAM queues is adjusted.

When a PCM page is hit, FWQ-LRU first computes *positive effects* and *negative effects*. If the page needs to be moved, FWQ-LRU performs a page migration. Then, the page is moved to the MRU position of the LRU list. When a page fault occurs, FWQ-LRU gets a free frame in the hybrid memory, puts the requested page in the frame, and inserts the page into the MRU position of the LRU list. If the page is placed in DRAM, the page should be inserted into a DRAM queue. If the request is a write request, FWQ-LRU counts the write requests and pushes the record of the write request into FWQ. Then, FWQ-LRU checks if the buffer serves a certain number of write requests that reach the adjustment cycle. If yes, FWQ-LRU calculates *Page_Migration_Benefit* and adjusts the FWQ size.

Algorithm 2. FWQ-LRU (*r*)
Input: a page request *r*
Output: a reference to the requested page *p*
Preliminary: (1) *L* is the LRU list of all pages in the hybrid memory.
(2) *p* is the request page.
(3) *FWQ* is the FIFO write queue

```
 1:  if p is in DRAM then
 2:  |    Effect_Count(p, r); /* count the positive/negative effect */
 3:  |    move p to the MRU position of L;
 4:  |    if r is a write request then
 5:  |    |    DRAM_Queue_Manage (p, r, false); /* adjust the DRAM queue */
 6:  else if p is in PCM then
 7:  |    Effect_Count(p, r);
 8:  |    if r is a write request and there is a record of p in FWQ then
 9   |    |    Page_Migration(p);
10:  |    move p to the MRU position of L;
11:  else
12:  |    s = Fix_Page(r);
13:  |    insert p into s;
14:  |    link p to the MRU position of L;
15:  |    if p is in DRAM then
16:  |    |    DRAM_Queue_Manage (p, r, true);
     /* Adjust the size of FWQ */
17:  if R is a write request then
18:  |    write_request_count++;
19:  |    push the record p into FWQ;
20:  if write_request_count >= adjustment_cycle then
21:  |    adjust the size of FWQ;
22:  return the reference to p;
```

4   Performance Evaluation

4.1   Settings

We conduct simulation experiments to evaluate our algorithm. We use DPHSim [8] to simulate the hybrid memory architecture, where the DRAM-to-PCM ratio is set to 1:9. Then, we implement all the algorithms based on the simulator. The parameters used in FWQ-LRU are shown in Table 2. The parameters used in Hybrid-LRU and APP-LRU are the same as those in the original papers [10, 12].

We perform experiments on both synthetic and real traces. Each trace consists of a series of page references, and each page reference is composed of a page number and the request type (read or write). The properties of these traces are shown in Table 2 and Table 3.

Synthetic Traces. We generate Zipf traces which simulate the Zipf distribution of page references. We finally generate five Zipf traces that simulate different workloads with different read/write ratios and localities. Each synthetic trace contains 500,000 page references which focus on 10,000 pages whose page numbers range from 0 to

9999. Each trace has a different read/write ratio and locality value. Here, the locality means the spatial locality of all page references. For example, *Zipf9182* has a read/write ratio of 90%/10% and a locality of 80%/20.

Real OLTP Trace [17]. The real trace we used in the experiments is a one-hour OLTP trace in a bank system, which contains 470,677 read references and 136,713 write references, focusing on 51,800 pages in a 20GB CODASYL database.

All the experimental results are collected in the following subsections after a warmup phase in the hybrid memory. The warmup phase is defined as the time period that the buffer becomes full by performing random page accesses.

Table 2. Parameters for FWQ-LRU

Parameter	Value
Initial FWQ size	The same as the DRAM size
max_hq	90% of the DRAM size
Adjustment cycle	Twice the memory size
Factor$_{incre}$	0.1
Factor$_{decre}$	0.8
Factor$_{axe}$	0.5

Table 3. Features about the synthetic and real traces

Trace	# of Requests	# of Pages	R/W Ratio	Locality
Zipf9182	500,000	10,000	90%/10%	80%/20%
Zipf9155	500,000	10,000	90%/10%	50%/50%
Zipf5582	500,000	10,000	50%/50%	80%/20%
Zipf1982	500,000	10,000	10%/90%	80% / 20%
Zipf1955	500,000	10,000	10%/90%	50%/50%
Real OLTP	607,390	51,880	77%/23%	–

4.2   Results

We compare the performance of FWQ-LRU with other algorithms under a small buffer. We vary the memory size from 20% to 60% of the total number of referenced pages in the trace for each trace. For example, in T9182, T9155, T5582, T1982, and T1955, the memory size varies from 2000 to 6,000 pages because there are totally 10,000 data pages referenced in each of these traces. Accordingly, the memory size for the OLTP trace ranges from 10376 to 31128 pages.

Page Fault Count. Figure 4 shows the page fault counts of LRU, FWQ-LRU, and Hybrid-LRU for the Zipf traces as well as the real OLTP trace. APP-LRU and MHR-LRU have the same hit ratio as LRU so we do not show them in the figure. FWQ-LRU also has the same hit ratio as LRU. Hybrid-LRU has many page faults, and it is much serious under large memory for write-intensive traces. Because the DRAM size is limited, extensive memory and intensive write accesses make more pages migrated from PCM to DRAM, so more DRAM page release occurs in Hybrid-LRU, which cause more page faults.

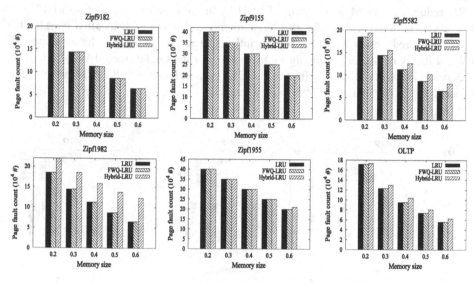

Fig. 4. Page faults

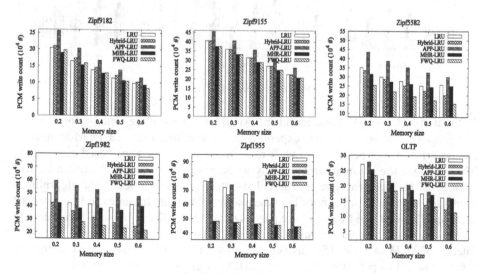

Fig. 5. PCM write counts

PCM Write Count. Figure 5 shows the total numbers of PCM writes of LRU, Hybrid-LRU, APP-LRU, MHR-LRU, and FWQ-LRU under the Zipf traces and the real OLTP trace. For each Zipf trace, we vary the memory size from 20% to 60% of the total number of different referenced pages, from 2000 to 6000 pages. The PCM write count of FWQ-LRU is lower than that of the other algorithms in most cases. FWQ-LRU reduces the PCM write count of LRU by an average of 23% and up to 47.3%. Compared with Hybrid- under the large memory size for high-locality workloads, FWQ-LRU reduces PCM writes significantly. The results show that the estimator of FWQ-LRU is helpful for reducing PCM writes. For the OLTP trace, FWQ-LRU also reduces PCM writes efficiently for the real OLTP trace. It reduces the PCM write count of LRU by 21.5% and up to 30.9% on average.

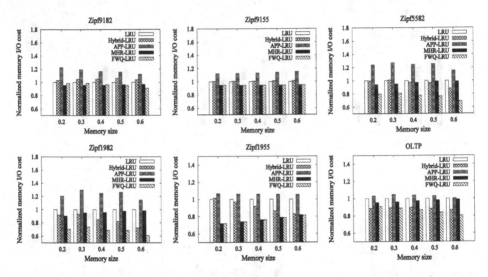

Fig. 6. Memory I/O costs

Estimated I/O Cost. In this experiment, we aim to measure the time performance of our proposal. However, as our experiments are conducted over a PCM-simulation framework, it is not possible to obtain real run time for each trace-based experiment. Therefore, we use an approximation way to estimate the overall time performance of each buffer management scheme. Particularly, the overall run time of each algorithm is computed as the sum of memory I/O time and disk I/O time. In our experiment, we set the *cacheline* size as 64 bits and the page size as 2 KB (32 *cachelines*). According to Table 1, we set the read latency of PCM to 1.6 μs (50 ns per *cacheline*), the write latency of PCM to 16 μs (500 ns per *cacheline*), the DRAM page-access latency to 1.6 μs and (50 ns per *cacheline*), and the HDD page-access latency to 5 *ms*.

Figure 6 shows the memory I/O costs of each algorithm, which are normalized according to LRU. As shown in the figure, FWQ-LRU has the best memory I/O performance for all workloads. This is because FWQ-LRU can reduce PCM writes and only introduces a few DRAM writes.

4.3   Adaptivity of FWQ-LRU

In this section, we use a hybrid trace that consists of five Zipf traces to reveal the adaptivity of FWQ-LRU. The hybrid trace is generated by appending T9155 to the end of T9182 then orderly appending T5582, T1982, T1955 to the end of the trace.

Figure 7 shows the page fault counts of Hybrid-LRU, FWQ-LRU, and LRU. Under the hybrid trace with various types of access patterns, FWQ-LRU keeps the same page fault count as LRU, and Hybrid-LRU still has a higher page fault than LRU and FWQ-LRU. Figure 8 shows PCM write counts of the algorithms for the hybrid trace. FWQ-LRU has the least PCM write count, and it reduces the average 28.6% PCM write count of LRU. The results show that FWQ-LRU can adapt to the changes of access patterns and efficiently reduce PCM writes. Figure 9 shows the memory I/O costs of each algorithm under the hybrid trace, which is normalized according to LRU. FWQ-LRU has the best memory I/O performance, showing the advantage of the adaptive page migration policy in FWQ-LRU.

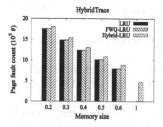
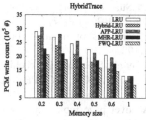
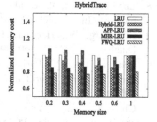

Fig. 7. Page fault counts under the hybrid trace

Fig. 8. PCM write counts under the hybrid trace

Fig. 9. Normalized memory I/O costs under the hybrid trace

5   Conclusions and Future Work

PCM has emerged as one of the most promising memories to be used in main memory hierarchy. This paper focuses on PCM/DRAM-based hybrid memory architecture and studies the buffer management scheme for such hybrid architecture. We propose a new buffering method called FWQ-LRU. Differing from previous studies, FWQ-LRU can adaptively adjust its page migration mechanism to gain more benefits on reducing PCM writes and improving the overall performance of the hybrid memory. We conducted trace-driven experiments and used both synthetic and real traces to compare our proposal with four existing schemes. The results show that FWQ-LRU can efficiently reduce PCM writes with a limited increase of DRAM writes while maintain a high hit ratio, yielding better overall performance for the hybrid memory architecture.

Acknowledgements. This study is supported by the National Science Foundation of China (62072419).

References

1. Liu, R., Jin, P., et al.: Efficient wear leveling for PCM/DRAM-based hybrid memory. In: HPCC, pp. 1979–1986 (2019)
2. Pozidis, H., Papandreou, N., et al.: Circuit and system-level aspects of phase change memory. IEEE Trans. Circuits Syst. II Express Briefs **68**(3), 844–850 (2021)
3. Jin, P., Wu, Z., et al.: A page-based storage framework for phase change memory. In: MSST, pp. 1–12 (2017)
4. Dhiman, G., Ayoub, R., et al.: PDRAM: a hybrid PRAM and DRAM main memory system. In: DAC, pp. 664–669 (2009)
5. Asadinia, M., Sarbazi-Azad, H.: The emerging phase change memory. Adv. Comput. **118**, 15–28 (2020)
6. Xia, F., Jiang, D., et al.: A survey of phase change memory systems. J. Comput. Sci. Technol. **30**(1), 121–144 (2015)
7. Fu, Y., Wu, Y.: CARAM: a content-aware hybrid PCM/DRAM main memory system framework. In: He, X., Shao, E., Tan, G. (eds.) NPC 2020. LNCS, vol. 12639, pp. 243–248. Springer, Cham (2021). https://doi.org/10.1007/978-3-030-79478-1_21
8. Zhang, D., Jin, P., et al.: DPHSim: a flexible simulator for DRAM/PCM-based hybrid memory. In: APWeb/WAIM, pp. 319–323 (2017)
9. Effelsberg, W., Haerder, T.: Principles of database buffer management. ACM Trans. Database Syst. **9**(4), 560–595 (1984)
10. Seok, H., Park, Y., Park, K.W., et al.: Efficient page caching algorithm with prediction and migration for a hybrid main memory. ACM SIGAPP Appl. Comput. Rev. **11**(4), 38–48 (2011)
11. Chen, K., Jin, P., Yue, L.: A novel page replacement algorithm for the hybrid memory architecture involving PCM and DRAM. In: Hsu, C.-H., Shi, X., Salapura, V. (eds.) NPC 2014. LNCS, vol. 8707, pp. 108–119. Springer, Heidelberg (2014). https://doi.org/10.1007/978-3-662-44917-2_10
12. Wu, Z., Jin, P., Yang, C., Yue, L.: APP-LRU: a new page replacement method for PCM/DRAM-based hybrid memory systems. In: Hsu, C.-H., Shi, X., Salapura, V. (eds.) NPC 2014. LNCS, vol. 8707, pp. 84–95. Springer, Heidelberg (2014). https://doi.org/10.1007/978-3-662-44917-2_8
13. Chen, K., Jin, P., Yue, L.: Efficient buffer management for PCM-enhanced hybrid memory architecture. In: APWeb, pp. 29–40 (2015)
14. Wang, H., Shen, Z., et al.: CLOCK-RWRF: a read-write-relative-frequency page replacement algorithm for PCM and DRAM of hybrid memory. In: HPCC, pp. 189–196 (2020)
15. Lee, S., Bahn, B., et al.: CLOCK-DWF: a write-history-aware page replacement algorithm for hybrid PCM and DRAM memory architectures. IEEE Trans. Comput. **63**(9), 2187–2220 (2014)
16. Lee, E., Bahn, H.: Caching strategies for high-performance storage media. ACM Trans. Storage **10**(3), 11:1–11:22 (2014)
17. Jin, P., Ou, Y., et al.: AD-LRU: an efficient buffer replacement algorithm for flash-based databases. Data Knowl. Eng. **72**, 83–102 (2012)

Efficiency-First Fault-Tolerant Replica Scheduling Strategy for Reliability Constrained Cloud Application

Yingxue Zhang[1,2], Guisheng Fan[1(✉)], Huiqun Yu[1], and Xingpeng Chen[1,2]

[1] Department of Computer Science and Engineering,
East China University of Science and Technology, Shanghai, China
{gsfan,yhq}@ecust.edu.cn
[2] Shanghai Key Laboratory of Computer Software Evaluating and Testing,
Shanghai, China

Abstract. Reliability requirement assurance is an important prerequisite for application execution in the cloud. Although copy management can improve the reliability of applications, it also brings a series of resource waste and overhead issues. Therefore, the efficiency-first fault-tolerant algorithm (EFFT) with minimum execution cost in the cloud application is proposed. This algorithm minimizes the execution cost of the application under the constraints of reliability, and solves the problem of excessive overhead caused by too many copies. The EFFT algorithm is divided into two stages: initial allocation and dynamic adjustment. On the initial allocation of EFFT algorithm, a sorting rule is defined to determine the priority of tasks and instances. During the adjustment phase, by defining an actual efficiency ratio indicator to measure the cost-effectiveness of an instance, the EFFT algorithm makes a good trade-off between cost and reliability in order to minimize execution costs. Run our algorithm on randomly generated parallel applications of different scales and compare the experimental results with four advanced algorithms. The experiments show that the performance of the algorithm we proposed is better than the other algorithms in terms of execution cost and fault tolerance.

Keywords: Cloud computing · Fault tolerance · Resource scheduling · Execution cost

1 Introduction

Cloud computing is a distributed computing model, which can perform computing tasks efficiently by deploying tasks in a resource pool with super computing power. The resources provided by cloud computing are highly scalable and pay-as-you-go. Companies and enterprises can obtain the services they need through cloud computing. However, there are many challenges in providing services through the cloud. Among them, software, human error, and hardware

© IFIP International Federation for Information Processing 2022
Published by Springer Nature Switzerland AG 2022
C. Cérin et al. (Eds.): NPC 2021, LNCS 13152, pp. 131–143, 2022.
https://doi.org/10.1007/978-3-030-93571-9_11

failures will affect the user's quality of service (QoS) significantly [1]. If the failure is not handled in time, it will cause huge losses to the user. Therefore, it is essential to improve the quality of service for users by providing fault-tolerant strategies.

Fault-tolerant scheduling is an effective method to improve the reliability of workflow. The existing fault-tolerant methods can be divided into two categories: active fault-tolerant and passive fault-tolerant [2]. Active fault tolerance generally uses strategies to avoid errors before failure occurs. These methods are characterized by the use of techniques such as monitoring, prediction and preemption. Passive fault tolerance is to take measures to reduce the impact of the failure after the failure occurs, such as replication, checkpointing, restarting, detection and recovery. The commonly used fault-tolerant technology is primary-backup replication, which means that a primary task will have zero, one, or multiple backup tasks. It is an important reliability enhancement mechanism. Too many replicas will cause a waste of resources, so how to reduce costs while ensuring reliability is also a big challenge.

Cloud providers usually have multiple data centers with different instance types and charge per hour, such as Amazon EC2 [3]. This paper is based on the hourly cost model for research. The characteristic of the hourly-based cost model is that if the resource is used for less than one hour, it is also charged at the hourly price. Considering this characteristic, we propose a new method to solve the problem of cost optimization under the constraint of reliability. The main contributions of this study are as follows.

- In order to solve the cost problem of reliability constrained cloud applications, We propose an efficiency-first fault-tolerant algorithm (EFFT) with minimum execution cost. In the initial assignment of tasks, the algorithm lists the instance types according to their efficiency ratio, and takes the most efficient instances constrained by the reliability.
- Taking into account the characteristics of the hourly cost model, we design a dynamic adjustment strategy to achieve the lowest cost. The algorithm defines an indicator of actual efficiency ratio to measure the cost performance of instances to assist in making a good trade-off between cost and reliability.
- Extensive experiments on randomly generated applications to evaluate the proposed algorithm and the four most advanced algorithms. Experimental results show that the EFFT algorithm can reduce waste of resources effectively, meet the constraints of target reliability, and minimize execution costs.

The rest of this article is organized as follows. The related works are summarized in Sect. 2. Section 3 describes the considered problem. The proposed EFFT algorithm is depicted in Sect. 4. Section 5 shows the performance evaluation of the proposed algorithm, followed by conclusions and future work in Sect. 6.

2   Related Work

This section reviews the related fault-tolerant scheduling of DAG-based work-flows. Many scholars have widely studied fault-tolerant scheduling in different situations.

Uncertainty in the cloud environment is the main reason that affects the quality of services, so we need to model and evaluate these uncertainties. Tang et al. [4] identified a set of potentially defective components based on the negative symptoms observed by the user. Each potentially faulty component was then evaluated to quantify its fault likelihood. Schatz and Wang [5] proposed a widely accepted reliability model, using Poisson distribution to describe the probability of virtual machine failure. This reliability model is widely used in this field of fault-tolerant scheduling [7–9].

In terms of fault-tolerant scheduling, Yao et al. [10] proposed a Hybrid Fault-Tolerant Scheduling Algorithm (HFTSA) for independent tasks with deadlines by integrating resubmission and replication techniques in virtualized cloud systems. In the process of task scheduling, HFTSA selects an appropriate fault-tolerant strategy for each task in resubmission and replication according to the characteristics of tasks and resources. A elastic resource allocation mechanism is designed to dynamically adjust the provided resources and improve resource utilization. In order to minimize the redundancy, Xie et al. [14] proposed the enough replication for redundancy minimization (ERRM) algorithm to satisfy reliability requirement. To overcome the time complexity of the ERRM algorithm and to handle large-scale parallel applications, they proposed the heuristic replication for redundancy minimization (HRRM) algorithm based on the ERRM algorithm. Considering that a minimum number of replicas does not necessarily lead to the minimum execution cost and shortest schedule length, Xie et al. [15] proposed the quantitative fault-tolerance with minimum execution cost algorithms QFEC and QFEC+ with minimum execution costs and quantitative fault-tolerance with shortest schedule length algorithms QFSL and QFSL+ to satisfy the reliability requirements of workflows. Samaneh et al. [16] proposed Improving Clusters with Replication (ICR) algorithm which minimizes the workflow execution cost under the constraints of deadline and reliability. This algorithm includes three stages: the Scheduling, the Fix Up, and the Task Replication. The most critical part of the algorithm is to use the rest of the idle time slots in leased resources to replicate tasks. The fault-tolerant scheduling types of workflow can be divided into two categories: strict scheduling and general scheduling [15]. Strict scheduling means that each task should be executed after all replicas of its parent task are executed, regardless of success or failure. General scheduling is that the current task starts immediately after any replica of the parent task is completed. In this paper, we only discuss the predictable strict scheduling mode.

The above works are based on the general cost model for research, without considering the actual situation. In fact, cloud service providers charge for cloud services at an hourly price. The above algorithm cannot well solve the problem of minimizing the cost in the hourly cost model, while considering the user's service quality.

3 Problem Description

The problem considered in this paper is how to minimize the cost under the constraint of reliability, so as to reduce the waste of resources and save the cost of users. In this section, we will introduce the system model, the related reliability model and hourly-based cost model. At the end of this section we give the problem definition.

3.1 System Model

The cloud can provide a set of k types of instances for users, which can be expressed as $\{v_1, v_2, ..., v_k\}$. The cloud environment can be modeled as a fully connected undirected graph $G_c = (V, E)$, where vertices $V = \{v_{i,k}\}$ indicates different number of instances of type k and E indicates the transmission path of the full connection . The processing power of the various instance types can vary. Usually cloud providers use a metric names Compute Unit(CU) to represent the processing capabilities of different types of instances. The higher the CU value, the stronger the processing power of this type of instance. In this paper, we use the parameter $speed_k$ to represent the processing speed.

A microservice-based application can be modeled as a directed acyclic graph (DAG) $G = (T, E)$. Each task of a microservice is represented by t_i, $T = \{t_1, t_2, ...t_n\}$, where n represents the number of subtasks. The dependency between tasks is represented by directed edges e_{ij} between corresponding nodes. $E = \{e_{ij}\} = \{(t_i, t_j) | t_i \epsilon P_i\}$, where P_i represents all parent nodes of task t_i. Applications are deployed on different types of instances in the form of microservices.

3.2 Reliability Model

Reliability is defined as the probability of successful execution of the task. This paper mainly considers transient faults, and transient faults during task operation follow the Poisson distribution [5]. We adopt λ_k to represent the failure rate of an instance of type k and $w_{i,k}$ to represent the execution time of task t_i running on instance v_k. Then the reliability of t_i in v_k is denoted by

$$R(t_i, v_k) = e^{-\lambda_k w_{i,k}} \tag{1}$$

In this paper, we use an active replication scheme to perform replica fault tolerance. Active replication will generate multiple replicas for each task to ensure reliability. A task t_i with m replicas can be denoted by $\{d_{i,1}, d_{i,2}, ..., d_{i,m}\}$. The reliability of t_i with m replicas is

$$R_{t_i} = 1 - \prod_{j=1}^{m} \left(1 - R_{t_i}^j\right) \tag{2}$$

The reliability of the application is the product of those of all tasks:

$$R(G) = \prod_{i=1}^{n} R_{t_i} \tag{3}$$

3.3   Cost Model

The cost model is a pay-as-you-go model, which is an hourly-based cost model. Each instance has its own charging standard, based on the different resources provided by it. We use $perCost_k$ to represent the unit time cost of the instance v_k, then the total cost of the application is the sum of the execution costs of all task copies, which is also equal to the multiplication of the total time running on the instance and the unit price of the execution cost, that is

$$cost(G) = \sum_{v_{i,k} \in V} cost(v_{i,k}) = \sum_{v_{i,k} \in V} \left(\sum_{j=1}^{num} (ET_j + TT_j) \times perCost_k \right) \tag{4}$$

Where ET represents the execution time of the task t_j on the i-th instance of type k $v_{i,k}$, and TT represents the transmission time.

3.4   Problem Statement

Given an application G with reliability requirements $R_{req}(G)$, each of its functions is implemented by microservices, and it is expressed as a DAG workflow form. There is a fully connected cloud computing environment that provides different types of instances for services. Usually an instance can be regarded as a virtual machine. The research problem is to find a scheduling scheme $X_i = \{\chi_1, \chi_2, ..., \chi_n\}$ with the least cost of resource consumption while meeting the reliability requirements, where each X represents the allocation of replicas of task t_i, $\chi_i = \{\{d_{i,1}, v_{1,k}\}, \{d_{i,2}, v_{2,k}\}, ..., \{d_{i,m}, v_{j,k}\}\}$.

Target: We aim to allocate replicas and corresponding instances for each task to reduce the cost of the application under reliability requirement $R_{req}(G)$. The formula is as follows:

$$\min \ Cost(G)$$
$$s.t. \ \ R(G) = \prod_{i=1}^{N} R_{ti} \geq R_{req}(G) \tag{5}$$

4   Algorithm Implementation

Dag-based fast execution workflow task scheduling is a well-known NP-hard optimization problem [6]. In this section, we propose a heuristic fault-tolerant algorithm based on the hourly cost model, which can find the least costly scheduling scheme under the constraints of reliability. The efficiency-first fault-tolerant algorithm (EFFT) algorithm first sorts the tasks and instances separately. Then perform initial scheduling based on the efficiency ratio of different types of instances, and finally adjust the allocation based on the actual efficiency ratio of each instance. In the following sections, we will elaborate on the details of the EFFT algorithm.

4.1 Task Prioritizing

As a classic algorithm, HEFT algorithm has very important guiding significance in solving workflow scheduling problems [13]. Inspired by HEFT, we modify the upward ranking value and target to minimize its resource consumption cost. The formula for the new upward sort value is defined as

$$rank_u\left(t_i\right) = \bar{\omega}_{i,k} + \max_{t_j \in succ(t_i)} \left\{\gamma_m \cdot e_{i,j} + rank_u\left(t_j\right)\right\} \tag{6}$$

Where $\bar{\omega}_{i,k}$ represents the average execution time of the task in the instance type k, $\bar{\omega}_{i,k} = \frac{\sum_{k=1..M} \gamma_k \cdot \omega_{i,k}}{M}$. We assign priority to tasks in descending order of $rank_u\left(t_i\right)$.

4.2 Instance Prioritizing

The processing capabilities of different types of instances may vary. Therefore, the execution time of the same task on different types of instances is different. We define the efficiency rate as a metric to measure the ratio of the processing speed to the price of an instance, which is defined as follows:

$$EffRate_k = \frac{speed_k}{perCost_k} \tag{7}$$

We sort the list in descending order according to the efficiency rate of the instance types. If there are multiple instance types with the same efficiency rate, the fastest one is prioritized.

4.3 Initial Allocation Scheme

Algorithm 1 gives the pseudo code of the initial allocation scheme. Firstly, make a lower bound judgment on the reliability of the tasks in the instance. If one task has $R\left(t_i, v_k\right) < R_{req}\left(G\right)$, then no matter how many replicas for any other tasks, $R_{req}(G)$ cannot be satisfied (Lines 2–3). Calculate the priority of tasks and instances, and assign the most efficient instance to each task. Note that the most efficient instance may not be the most reliable. So every time a task is assigned, we update the current reliability value of each task (Lines 4–11). Calculate the current reliability value of the application (Line 12). If the previous scheduling scheme meets the reliability constraints, skip the following steps. Otherwise, we need to calculate and allocate copies of the task. According to the priority order of the instances, the copies of the tasks are increased in a loop until the reliability constraints are met (Lines 13–25). Note that copies of the same task cannot be placed on the same instance.

Algorithm 1: Initial Scheduling Scheme

Input: Bag of Tasks: $G = (T, E)$, $R_{req}(G)$, Different types of instances:
$V = \{v_{i,k}\}$
Output: $X_i = \{\chi_1, \chi_2, ..., \chi_n\}$, $Cost(G)$

1 **begin**
2 $\quad$ **if** $\forall v_k \in V, \ \exists t_i \in T, \ R(t_i, v_k) < R_{req}(G)$ **then**
3 $\quad\quad$ return;
4 $\quad$ **for** $i = 1$ *to* n **do**
5 $\quad\quad$ **for** $j = 1$ *to* k **do**
6 $\quad\quad\quad$ **if** $R(t_i, v_j) < R_{req}(G)$ **then**
7 $\quad\quad\quad\quad$ $\chi_i \leftarrow v_{rank[j]}$;
8 $\quad\quad\quad\quad$ Update *taskList* and *instanceList*;
9 $\quad\quad\quad\quad$ break;
10 $\quad\quad$ **end**
11 $\quad$ **end**
12 $\quad$ Calculate $R(G)$ using Eq. (3);
13 $\quad$ **while** $R(G) < R_{req}(G)$ **do**
14 $\quad\quad$ **for** $i = 1$ *to* n **do**
15 $\quad\quad\quad$ **for** $j = 1$ *to* k **do**
16 $\quad\quad\quad\quad$ **if** *there are duplicate tasks on* v_j **then**
17 $\quad\quad\quad\quad\quad$ break;
18 $\quad\quad\quad\quad$ **else**
19 $\quad\quad\quad\quad\quad$ $\chi_i \leftarrow v_{rank[j]}$;
20 $\quad\quad\quad$ **end**
21 $\quad\quad$ **end**
22 $\quad\quad$ Updare X;
23 $\quad$ **end**
24 $\quad$ return X;
25 **end**

4.4 Dynamic Adjustment Scheme

In Algorithm 1, tasks are initially allocated. Since the cost model is based on hours, the result of the initial allocation plan may not be optimal, and some resources will be wasted. Therefore, we adjust the distribution plan in this part. The actual efficiency ratio of the instance is defined as follows:

$$EffRate_{actual,k} = \frac{speed_k}{actualCost_k} = \frac{speed_k \times ET_k}{perCost_k} \tag{8}$$

We reorder the instances according to the Eq. (8). If the priority is lowered, we make a migration plan for the tasks on the instance. The migration standard are as follows:

- The reliability of the target instance of the migration should be more than the lower bound of the required reliability of the task, the reason is explained in Section;

- There is no replica of the task on the target instance of migration, that is, a task cannot be assigned to the same instance repeatedly;
- The cost of the solution after the migration should be less than the solution before the migration.

After meeting the migration criteria, we update the task allocation on each instance and recalculate time and cost consumption. It should be noted that task redistribution not only changes its own cost, but also changes the cost of other tasks, because task redistribution changes the communication overhead of other tasks. Therefore, the cost and execution time of all tasks should be updated for each allocation until the final allocation plan is found. Algorithm 2 gives the pseudo code of the adjustment scheme. Firstly, we count the usage of the current instances and record all adjustable instances in a set, which is represented by the symbol *instanceList*. Then traverse the *instanceList* and calculate the actual efficiency ratio of the instances (Lines 2–5). Judge whether the task meets the migration conditions, and if so, update the plan until the least cost plan is found (Lines 7–13).

4.5   Complexity Analysis

The time complexity of the Efficiency-First Fault-Tolerant (EFFT) algorithm is analyzed as follows: Calculating the reliability of the application must traverse all tasks, which can be done within $O(|N|)$ time. The total number of replicas for each task must be lower or equal to the number of instances m, so the time complexity is $O(|M|)$. All task and instance type need to be traversed during the initial allocation, so the time complexity is $O(|N| \times |K|)$. When dynamically adjusting the scheme, it is necessary to traverse all the replicas u and used instances m, so the time complexity is $O(|U| \times |M|)$. Therefore, the total time complexity of the EFFT algorithm is $O(|U| \times |M|)$.

5   Performance Evaluation

In this section, we provide a discussion on the simulation results. We will schedule many randomly generated applications by comparing the proposed EFFT algorithm with the other four state-of-the-art methods to show the experimental results. The four most advanced methods are MaxRe [11], RR [13], ERRM [14], and QFEC [15]. We use the Pegasus Project Synthesis Workflow Generator to randomly generate simulation applications with different structures. To evaluate the scalability of the algorithm, applications with the number of subtasks of 40, 45, 50, 55...100 are considered for our experiments. In each problem size, we randomly generate 50 workflows. We employ a pricing model for instances according to real-world public cloud providers such as Amazon EC2, and this pricing model provides a reference for the costing of the assigned application. Price is a linear function of the number of processing units in different instance types and is charged per hour. The faster the processing speed, the higher the hourly cost.

Algorithm 2: Dynamic Adjustment Scheme

Input: Bag of Tasks: $G = (T, E)$, $R_{req}(G)$, Different types of instances:
$V = \{v_{i,k}\}$, $X_i = \{\chi_1, \chi_2, ..., \chi_n\}$, $Cost(G)$
Output: $X_i' = \{\chi_1, \chi_2, ..., \chi_n\}$, $Cost'(G)$

1 **begin**
2   Calculate the usage of instances $instanceList$
3   **while** $Cost'(G) < Cost(G)$ **do**
4   **for** $i : instanceList$ **do**
5   Calculate $EffRate_{actual,i}$ using Eq. (8);
6   Calculate new priority and compare;
7   Record a list of adjustable instances $instanceList_{adj}$;
8   **for** $j : instanceList_{adj}$ **do**
9   **if** *tasks on the instance can be migrated* **then**
10   Migration task from i to j;
11   Update X;
12   **end**
13   **end**
14   Calculate execution time and $Cost'(G)$;
15   **end**
16 **end**

In this paper, 15 different types of instances are used. The failure parameters of each type of instance are: 0.000005, 0.000006, 0.000007,....,0.000020. With the increase of computing power, the failure rate of the instance will become higher.

5.1   Benchmark Algorithm

In this section, we will introduce four benchmark algorithms.

The reliability of the application is calculated by multiplying the reliability values of all tasks, as shown in Eq. 3. The MaxRe algorithm solves this problem by transferring application's reliability requirement to the sub-reliability requirements of tasks [11]. The sub-reliability requirement for each task is calculated by

$$R_{req}(t_i) = \sqrt[n]{R_{req}(G)} \tag{9}$$

Firstly, the instances are sorted according to the reliability value. Then select the most reliable instance to assign tasks and replicas until the reliability requirements of each task are met.

The main limitation of the MaxRe algorithm is that the sub-reliability requirements of all tasks are equal and high, so there are more replicas of redundancy. The RR algorithm reduces redundancy [13]. By changing the reliability calculation method of non-root tasks. The sub-reliability requirements in the RR algorithm are calculated continuously based on the actual reliability achieved by previous allocations:

$$R_{req}\left(t_{seq(i)}\right) = \sqrt[n-i+1]{\frac{R_{req}\left(G\right)}{\prod_{x=1}^{i-1} r\left(t_{seq(x)}\right)}} \tag{10}$$

It also selects the most reliable instance to allocate tasks and replicas until the reliability requirements of each task are met.

The ERRM algorithm is the enough replication for redundancy minimization algorithm proposed by Xie [14]. The lower bound of algorithm redundancy is equal to the reliability value of the given target. Each task iteratively selects the processor with the largest reliability value until the lower bound of the task's reliability requirements is met. Then, it's until application's reliability requirement is satisfied that the available replicas and correspond processors with the maximum reliability values are selected iteratively.

Choosing the most reliable instance can only ensure that the number of used instances is as small as possible, but does not guarantee that the cost is the lowest. Similarly, the minimum number of deplicas does not necessarily bring the lowest cost. The quantitative fault-tolerance with minimum execution cost(QFEC) algorithm defines the reliability of all unassigned subtasks as the upper bound [15], then the reliability value of the assigned task is as follows:

$$R_{req}\left(t_{seq(i)}\right) = \frac{R_{req}\left(G\right)}{\prod_{x=1}^{i-1} R\left(t_{seq(x)}\right) \times \prod_{y=i+1}^{n} R_{up}\left(t_{seq(y)}\right)} \tag{11}$$

Finally, iteratively select the instance with the smallest execution time for each task until the sub-reliability requirements are met.

5.2   Experimental Results and Discussion

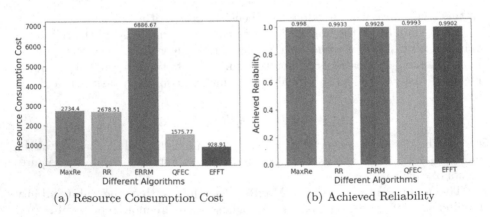

(a) Resource Consumption Cost   (b) Achieved Reliability

Fig. 1. Results of compared approaches on the randomly generated application

Figure 1 shows the cost and reliability of different algorithms. In this experiment, the target reliability value is set to 0.990, and the application program with a task number of 55 is used. We run each algorithm 100 times and use the average of their results as the final result. The experimental results are shown in Fig. 1. Figure 1(a) shows that our proposed approach has the least resource consumption cost that is even better than the quantitative fault-tolerance with minimum execution cost (QFEC) algorithm. This is because the QFEC algorithm does not take into account the characteristics of the hour-based cost model, but selects the fastest instance to handle the task. In the EFFT algorithm, we define the efficiency ratio and finally achieved the optimal cost. The cost of MaxRe and RR algorithm is higher than EFFT algorithm and QFEC algorithm. This is because these two algorithms prefer to select the most reliable instance to run the task, but the most reliable instance may not have high cost performance. Figure 1(b) shows that the reliability achieved by the MaxRe and RR algorithms are better than the EFFT algorithm we proposed. All five algorithms meet reliability constraints.

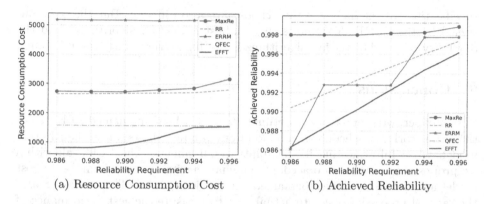

(a) Resource Consumption Cost   (b) Achieved Reliability

Fig. 2. Results of compared approaches with different reliability requirements

Figure 2 shows the influence of different reliability on the experimental results. The target reliability ranges from 0.988 to 0.998, and the incremental step size is 0.002. The results in Fig. 2(a) show that with the increase in reliability, the cost of our proposed algorithm is still optimal. Figure 2(b) shows that the cost of all algorithms has an approximately linear growth relationship with the reliability requirements. It is reasonable that as the reliability increases, the number of copies of the task will also increase, so that the resource consumption cost increases in an approximately linear manner.

Figure 3 shows the influence of different tasks on the experimental results under the same reliability. We set the target reliability value to 0.990, and investigate the effect of application level by varying it from 50 to 100. The number of tasks is 50, 55, 60, ... 100, respectively. From Fig. 3(a), we can find that as the level of the application increases, the cost does not increase linearly. This is because different applications consume different amounts of resources, and the

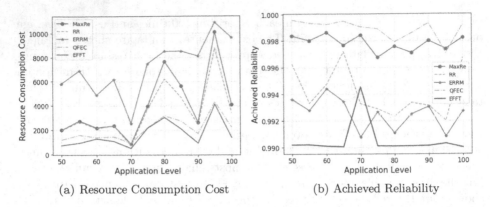

(a) Resource Consumption Cost   (b) Achieved Reliability

Fig. 3. Results of compared approaches with different reliability requirements

specifications of the instances they choose are also different. But the cost consumed by the EFFT algorithm is the smallest among all algorithms. And the reliability achieved by the five algorithms is also higher than the target reliability.

6   Conclusions

In this paper, we propose a cost-optimal resource scheduling named EFFT algorithm for workflow scheduling in a cloud environment based on an hourly cost model, while fully meeting its reliability requirements. The EFFT algorithm we proposed fully takes into account the characteristics of the hour-based cost model and defines an indicator named the efficiency ratio. This indicator uses the ratio of processing speed to actual cost to measure the cost performance of instances to assist in making a good trade-off between cost and reliability, so the algorithm can choose a better migration plan to adjust. The advantage of the EFFT algorithm is that the reliability it achieves is very close to the required reliability, thereby avoiding unnecessary excessive reliability. And the algorithm can be adjusted and migrated according to the actual running situation of the instance, so as to find the best solution to minimize the cost.

We compare our algorithm with four different baseline algorithms that are considered the most advanced. The experimental results prove that the EFFT algorithm has shown superior performance in dealing with various applications. In the future, we will improve this method to meet different types of needs, and will consider more attributes such as task type and response time.

Acknowledgements. This work was supported by the National Natural Science Foundation of China (No. 61772200), Shanghai Natural Science Foundation (No. 21ZR1416300).

References

1. Kalra, M., Singh, S.: Multi-objective energy aware scheduling of deadline constrained workflows in clouds using hybrid approach. Wireless Pers. Commun. **116**(3), 1743–1764 (2020). https://doi.org/10.1007/s11277-020-07759-4
2. Mukwevho, M.A., Celik, T.: Toward a smart cloud: a review of fault- tolerance methods in cloud systems. IEEE Trans. Serv. Comput. (2018)
3. Faragardi H R, Sedghpour M S, Fazliahmadi S, et al.: GRP-HEFT: A Budget-Constrained Resource Provisioning Scheme for Workflow Scheduling in IaaS Clouds **31**(6), 1239–1254 (2020)
4. Tang, Y., Shaer, E., Joshi, K.: Reasoning under uncertainty for overlay fault diagnosis. IEEE Trans. Network Serv. Manage. **9**(1), 34–47 (2012)
5. Shatz, S.M., Wang, J.P.: Models and algorithms for reliability-oriented task-allocation in redundant distributed-computer systems. IEEE Trans. Rel. **38**(1), 16–27 (1989)
6. Li, J., Liang, W., Huang, M., et al.: Reliability-aware network service provisioning in mobile edge-cloud networks. IEEE Trans. Parallel Distrib. Syst. **31**(7), 1545–1558 (2020)
7. Kumar, N., Mayank, J., Mondal, A.: Reliability aware energy optimized scheduling of non-preemptive periodic real-time tasks on heterogeneous multiprocessor system. IEEE Trans. Parallel Distrib. Syst. **31**(4), 871–885 (2020)
8. Kherraf, N., Sharafeddine, S., Assi, C.M., et al.: Latency and reliability-aware workload assignment in IoT networks with mobile edge clouds. IEEE Trans. Netw. Serv. Manage. **16**(99), 1435–1449 (2019)
9. Xie, G., Wei, Y.H., Le, Y., et al.: Redundancy minimization and cost reduction for workflows with reliability requirements in cloud-based services. IEEE Trans. Cloud Comput., 99 (2019)
10. Yao, G., Ren, Q., Li, X., Zhao, S.: Rub.: A hybrid fault-tolerant scheduling for deadline-constrained tasks in cloud system. IEEE Trans. Serv. Comput. (2020)
11. Zhao, L., Ren, Y., Xiang, Y., Sakurai, K.: Fault-tolerant scheduling with dynamic number of replicas in heterogeneous systems. In: Proceedings of the 12th IEEE International Conference on High Performance Computing and Communications, pp. 434–441 (2010)
12. Zhao, L., Ren, Y., Sakurai, K.: Reliable workflow scheduling with less resource redundancy. Parallel Comput. **39**(10), 567–585 (2013)
13. Hu, B., Cao, Z.: Minimizing resource consumption cost of DAG applications with reliability requirement on heterogeneous processor systems. IEEE Trans. Industr. Inf. **16**(12), 7437–7447 (2020)
14. Xie, G., Zeng, G., Chen, Y., et al.: Minimizing redundancy to satisfy reliability requirement for a parallel application on heterogeneous service-oriented systems. IEEE Trans. Serv. Comput. **13**(5), 871–886 (2020)
15. Xie, G., Zeng, G., Li, R.: Quantitative fault-tolerance for reliable workflows on heterogeneous iaas clouds. IEEE Trans. Cloud Comput. **8**(4), 1223–1236 (2020)
16. Nik, S.S.M., Naghibzadeh, M., Sedaghat, Y.: Task replication to improve the reliability of running workflows on the cloud. Clust. Comput. **24**(1), 343–359 (2021)

Towards an Optimized Containerization of HPC Job Schedulers Based on Namespaces

Tarek Menouer[1(✉)], Nicolas Greneche[1,2], Christophe Cérin[2], and Patrice Darmon[1]

[1] Umanis Research & Innovation, 92300 Levallois-Perret, France
{tmenouer,pdarmon}@umanis.com
[2] University of Paris 13, LIPN - UMR CNRS 7030, 93430 Villetaneuse, France
{nicolas.greneche,christophe.cerin}@univ-paris13.fr

Abstract. Recently, container technology is gaining increasing attention and has become an alternative to the traditional virtual machines artifact. The technology is used to deploy large-scale applications in several areas such as Big Data, AI, and High-Performance Computing (HPC). In the HPC field, several management tools exist as Slurm in one hand. On the other hand, the literature has considered many container scheduling strategies. The majority of container scheduling strategies don't think about the amount of data transmitted between containers. This paper presents a new container scheduling strategy that automatically groups containers that belong to the same group (Namespace) on the same node. In brief, the plan is application-aware as long as someone knows which containers should be grouped in the same Namespace. The objective is to compact the nodes with containers of the same group to reduce the number of nodes used, the communication inter-node costs, and improve containerized applications' overall Quality-of-Service (QoS). Our proposed strategy is implemented under the Kubernetes framework. Experiments demonstrate the potential of our strategy under different scenarios. Most importantly, we show first that cohabitation between our new scheduling strategy and the default Kubernetes strategy is possible and for the benefit of the system. Thanks to Namespaces, cohabitation is not limited to two methods for scheduling either batch jobs or online services. Second, thanks to the deployment automation, we also demonstrate that multiple Slurm clusters can be instantiated from a pool of bare metal nodes. This reality contributes to the concept of "HPC as a Service."

Keywords: Containers and virtualization · HPC and Big Data workloads · Scheduling and resource management · Optimization

C. Cérin et al. (Eds.): NPC 2021, LNCS 13152, pp. 144–156, 2022.
https://doi.org/10.1007/978-3-030-93571-9_12

1   Introduction

A container is a lightweight OS-level virtualization technique that enables the deployment and the execution of applications and their dependencies in a resource-isolated process on the cloud, edge/fog and Internet-of-Things platforms. The container technology is used to deploy large scale applications in challenging fields, such as big data analytics, High-Performance Computing (HPC), Scientific computing, Edge computing, and Internet-of-Things (IoT) [3].

In the field of HPC, the use of containers has experienced strong growth. The reason is that containers offer an easy way to deploy HPC jobs. However, the deployment of HPC jobs with containers poses new challenges concerning performance that need to be guaranteed for stream processing, batch processing applications, training, and testing environments.

In the literature, several container management tools are proposed to automate the management and scalability of applications as Docker SwarmKit [20], Apache Mesos [19], and Google Kubernetes [14,20]. The container management tools allow automatic scalability and deployment of applications on cloud infrastructure. The deployment of applications is carried out with a container scheduling strategy. The principle selects the most suitable node from a set of nodes that form the cloud platform to execute each container.

In this paper, we present a new container scheduling strategy validated in the context of the HPC ecosystem. Our motivation and approaches aim to improve the performance of deployment HPC jobs workflow in a cloud infrastructure. The idea for the improvement is to group all containerized jobs that communicate with each other on the same node. The approach can also reduce the number of cloud nodes used. Moreover, containers can cohabit with different container scheduling strategies in supplement with our policy. Among the challenges of the Cloud computing and HPC fields that we address in this paper, we would like to mention performance controlling and better resource utilization. We generally get better time performance when communication resides on the same node, hence our grouping strategy. Regarding resource utilization, the consolidation of servers [9] is also a well-proven energy conservation technique, for example, by taking advantage of the built-in power-saving features of modern hardware to reduce the energy consumption of servers. Summarizing, consolidation, and performances are the reason and importance of considering the association between containers.

For that challenges, we rely on the concept of Namespaces. Otherwise, if there is no room for all jobs, our strategy executes containerized jobs in the node with less load in running containers. Our approach aims to: (i) minimize the number of nodes used; (ii) reduce the communication overhead, and (iii) improve the overall Quality-of-Service (QoS) of interprocess communication of containerized HPC jobs.

In our study, the notion of a Namespace can be interpreted in several ways. For example, we can group the containers of the same application in the same Namespace. Or we can group the containers of the same user that belong to one or more applications in the same Namespace. Our study assumes that an expert knows the pertinent grouping to satisfy, and we offer a means to group containers.

Our approach is implemented with Go language implementation inside the Kubernetes framework[1], with a minimal change in the original Kubernetes code since we only need to add a new statement in the source code of Kubernetes. Kubernetes is a popular open-source framework widely used in the industry and academia. Thus, the contributions of the paper are:

– Propose a new container scheduling strategy allowing to reduce the cost of communication between containers belong to the same group (application/user);
– Extend our previous study proposed in [4] about the containerization of HPC jobs using our new container scheduling strategy and demonstrate how to deploy multiple Slurm clusters inside a pool of bare-metal machines.

The organization of the paper is as follows. Section 2 introduces some related works. Section 3 describes the principle of our approach. Section 4 presents how the HPC workflow is encapsulated in containers. Section 5 introduces exhaustive experiences that allow the validation of our proposed approach. At last, we introduce future works in Sect. 6.

2   Related Work

In this section, we will start by presenting in Subsect. 2.1 some studies proposed in the context of container scheduling. Then, we introduce in Subsect. 2.2 some studies that relate to containerize HPC jobs schedulers. Finally, we conclude our purpose by a positioning statement in Subsect. 2.3.

2.1   Container Scheduling Studies

In the containers' scheduling context several studies are proposed [6,8,10,11, 15,17].

In [11], authors propose a power-efficiency container scheduling approach based on a machine learning technique for the cloud computing environment. The novelty of the proposed approach is to reduce the global power consumption of heterogeneous cloud infrastructure. The principle is based on learning and scheduling steps applied each time a new container is submitted. The learning step is applied to estimate the power consumption of each node. Then it groups the nodes that form a heterogeneous cloud infrastructure into clusters according to their power consumption. The scheduling step is applied to select the node which has the lowest power consumption. The author's approach is implemented in Docker SwarmKit [20].

In [5], authors propose a dynamic allocation algorithm for Docker container resource, which adapts to applications' requirements. This study considers the

---

[1] See https://golang.org and https://kubernetes.io/.

characteristics of Docker containers, the needs of multiple applications, and the available physical resources. The goal is to give a resource allocation program that could make the cost of deployment minimum under the premise of getting a certain quality of service. That algorithm considers the resources on nodes, the network consumption of nodes, and the energy consumption of nodes. But it also did not consider the relationship between containers, nodes, and the internal relations among containers.

In [6], authors propose a study to evaluate how containers can affect the overall performance of applications in Fog Nodes. They analyze different container orchestration frameworks and how they meet Fog requirements to run applications. The authors also propose a container orchestration framework for fog computing infrastructures. In the same context, in [12], the authors propose a study to review the suitability of container technology for edge clouds and similar settings. It starts with summarising the virtualization principles behind containers, and then it identifies vital technical requirements of edge cloud architectures.

2.2 Studies Related to the Containerization in the HPC Ecosystem

In [2], authors studied the potential use of Kubernetes on HPC infrastructure for use by the scientific community. They directly compared both its features and performance against Docker Swarm and bare-metal execution of HPC applications. They made some hypotheses regarding the accounting for operations such as (1) underlying device access, (2) inter-container communication across different hosts, and (3) configuration limitations. They discovered some rules that showed that Kubernetes presents overheads for several HPC applications over TCP/IP protocol. Our work also considers this system-level, but we consider the native container support of Linux rather than a dedicated (Kubernetes) overlay.

In [1], authors use Kubernetes and Slurm (HPC jobs scheduler) to build an infrastructure with two scheduling stages. Their architecture is composed of one Kubernetes orchestrator and several Slurm HPC clusters. The user submits a custom kind of job called SlurmJob to Kubernetes orchestrator. That is when the first stage occurs. Kubernetes selects one of the Slurm driven HPC clusters according to resources requested by the user. When the job reaches the HPC cluster, Slurm schedules the job. This architecture is very similar to Grids [13] with a higher level of elasticity. The job itself is a singularity container.

In [18], authors describe a plugin named Torque-Operator. The proposed plugin serves as a bridge between the HPC workload manager Torque and the container orchestrator Kubernetes. Authors also propose a testbed architecture composed of an HPC cluster and a big data cluster where Torque-Operator enables scheduling container jobs from the big data cluster to the HPC cluster.

2.3 Positioning

The key difference between all of the cited works and our work is adopting a consolidation strategy, a technique inherited from the cloud world, for the HPC world. The consolidation strategy aims to improve performance in the sense of reducing communications between "tasks." The idea consists of grouping the containers belonging to the same Namespace in the same compute node. To summarize, our strategy is application-aware in the sense that if someone (a domain expert) knows how to label applications with Namespaces, then we will find a placement that reduces communication between containers. Our method is validated by deploying HPC workflow encapsulated in containers and managed by the Slurm tool. In [4], we have proposed a study that uses containers technology to instantiate a tailored HPC environment based on the user's reservation constraints. We claim that the introduction and use of containers in HPC job schedulers allow better management of resources more economically. From the use case of SLURM, we release a methodology for 'containerization' of HPC jobs schedulers which is pervasive, i.e., spreading widely throughout any layers of job schedulers.

3   Description of Our Container Scheduling Strategy

The goal of our approach is to give answers to the problem stated as follows: *in a cloud computing environment, how to minimize the number of used cloud nodes and also reduce the cost of communication between containers that belong to the same application/user?*

3.1   Problem Modeling

The problem study in this paper can be modelled as follows. Given a set of n nodes forming the cloud infrastructure: $node_1, node_2, \cdots, node_n$. Given also a set of k containers that belong to x namespace with $u \in [1 \cdots x]$. Goals are:

- *Minimize* the number of cloud nodes used to execute k containers,
- *Minimize* the cost of communication between containers.

The mathematical modeling of our problem can be define using two variables: (i) x_i; and (ii) $cost_{u,i,j}$.

$$x_i = \begin{cases} 1 & \text{If the node i is used} \\ 0 & \text{Otherwise} \end{cases}$$

$cost_{u,i,j}$ represents the cost of communication between $container_{u,i}$ and $container_{u,j}$ which belongs to the namespace u.

Thus, the objective functions are:

$$Min(\sum_{i=1}^{n} x_i) \wedge \tag{1}$$

$$Min(\sum_{u=1}^{x} cost(u, i, j), \quad \forall i, j \in \{1, \cdots, k\} \text{ with } i \neq j) \tag{2}$$

3.2   Principle of Our Container Scheduling Strategy

We now detail the heuristic we implemented to solve the mathematical problem. Our container scheduling strategy's primary purpose is to reduce the number of cloud nodes used and the cost of communication between containers deployed to execute the same jobs of the same requests. Each time a new container is submitted, our approach selects the node from all cloud infrastructure nodes with the most significant number of containers belonging to the same application to reduce communication costs. Our heuristic policy regarding the general problem also compacts nodes with containers to reduce the number of cloud nodes used.

According to the principle mentioned above, our approach is designed for two categories of users:

- User that has a set of containers belongs to the same application, in this case, our strategy groups containers in the same node to reduce the cost of communication between containers;
- Set of users such as each user has a set of containers belongs to the same application, in this case, our strategy compact the cloud nodes with containers of the same user. For example, if we have two users, and each one has three containers, our strategy place the three containers of the user 1 in node 1, then the three containers of user 2 will be placed in node 2. This distribution allows reducing the cost of communication between containers of each user.

Algorithm 1 Main loop of our container scheduling strategy

Require: n: number of nodes that form the cloud infrastructure
Require: c_u: new container that belongs to the Namespace u
1: NCN_u = Number of Containers belongs to the Namespace u in $node_1$
2: $NCDN_u$ = Number of Containers Does not belongs to the Namespace u in $node_1$
3: $Selected_{node} = node_1$
4: cmpt =0
5: **for** i=1; i ¡ n; i++ **do**
6:    **if** NCN_u > Number of Containers belongs to the Namespace u in $node_i$ **then**
7:       NCN_u = Number of Containers belongs to the Namespace u in $node_i$
8:       $Selected_{node} = node_i$
9:       cmp++
10:   **end if**
11: **end for**
12: **if** cmpt == 0 **then**
13:    **for** i=1; i<n; i++ **do**
14:       **if** $NCDN_u$ < Number of Containers Does not belongs to the Namespace u in $node_i$ **then**
15:          $NCDN_u$ = Number of Containers Does not belongs to the Namespace u in $node_i$
16:          $Selected_{node} = node_i$
17:       **end if**
18:    **end for**
19: **end if**
20: Execute c_u on $Selected_{node}$

As presented in Algorithm 1, for each new submitted container c_u, our strategy checks in all nodes of the cloud infrastructure the number of running containers that belong to the same Namespace (group) as c_u. Then, select the node with a significant number of running containers to reduce communication costs between containers of the same Namespace. Otherwise, if no node has at least one container with the same Namespace as c_u, in this case, our strategy selects the node with the smallest number of running containers in hoping to have in future new containers that have the same Namespace as c_u to assemble them.

4   HPC Workflow Encapsulated in Containers

The approach proposed in this paper aims to optimize our previous work introduced in [4] on HPC containers. The idea is to use containers technology to instantiate a tailored HPC environment based on the user's reservation constraints. From the use case, we release a methodology for containerization of Slurm HPC jobs scheduler, which is pervasive, i.e., spreading widely throughout any layers of job schedulers.

The containerization of Slurm is realized on top of the Kubernetes framework. Kubernetes is a Cloud orchestrator that dispatch pods on nodes (virtual or physical machines). A pod is a set of containers that must be located on the same node. Slurm is a job scheduler designed for HPC applications. It is divided into two main programs:

- slurmctld is the brain of the HPC cluster. It knows the topology of the cluster and handles HPC job scheduling.
- slurmd is the worker part of the HPC cluster. It receives HPC jobs from Slurmctld and forks a user-owned process to run the HPC job on selected node(s).

As we can observe, both Kubernetes and Slurm rely on nodes to run containers or HPC jobs. To avoid confusion between Kubernetes and Slurm nodes, we will refer to the first as Cloud nodes and the second as HPC nodes. In traditional HPC clusters, Slurmctld is installed on an administration node, and each HPC node has its instance of Slurmd. As shown in Fig. 1, a user submits a job to Slurmctld, then Slurmctld dispatch it to Slurmd instances on HPC nodes. This mode implies communication between Slurmctld and Slurmd processes. Each message is authenticated through a local Munge daemon running on each node. Consequently, each Slurm pod embeds two containers: one for Slurmd or Slurmctld and one for Munge.

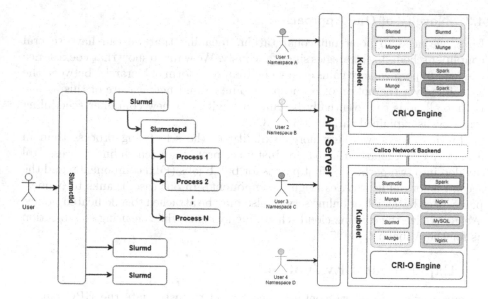

Fig. 1. Slurm architecture   **Fig. 2.** Containerized Slurm integration in Kubernetes

4.1 Containerization of Slurmctld

The pod is made of two containers, one for Slurmctld and another for Munged. Munged is required to authenticate communications between Slurmd workers and the slurmctld master. Slurmctld container talks to Munged container through a local UNIX socket located on an emptyDir which is a shared volume between the two containers.

Note that the Munged container is optional. If the pods are running on a dedicated Kubernetes Namespace with a network isolation through Network Policies, Munged can be disabled. If Munge is enabled, all the daemons of our architecture must share the same private key (eventually shared on a ReadOnlyMany capable Physical Volume).

4.2 Containerization of Slurmd

Slurmd is more tricky to containerize because of the requirements of resources limitation and security. By default, a container has no restrictions on resource consumption on the host. Cgroups enforce these limitations. The container engine instantiates slurmd within a Cgroups and configures subsystems to match the resources required by the user. In the Slurmd manifest, the number of CPUs used must be specified. This directive must check the number of cores declared in the Slurmd configuration in the container simulating the HPC node. Therefore, the use of an unprivileged container engine, i.e., launched by a non-root user, is impossible. As a consequence, the use of a privileged containers engine implies security issues.

4.3   Novelty of Our Approach

First, we need to point out cohabitation, meaning that we can have several container scheduling strategies simultaneously. We want to show that coexistence can lead to performance gains even when we have a form of hierarchy between the containers via the notion of Namespace. The second new feature of this article is that all pods can eventually be run only with our new container scheduling strategy, as explained earlier in Sect. 3.2.

In these ways, we offer more flexibility in the scheduling process than in a majority of Clouds. Alibaba, for instance, provides a scheduling process [16] divided into two parts. The first part is for batch jobs (Fuxi component), and the second is for online services (Sigma component). In our case, thanks to Namespaces, each batch job or online service also may be attached to a dedicated policy. We generalize the existing cloud scheduling systems while ensuring cohabitation as for the Alibaba cloud [7].

5   Experimental Evaluation

In our previous work, we containerized Slurm to instantiate the HPC cluster over a Kubernetes cluster. We now want to mix Slurmd pods with regular pods (like Nginx, MySQL, etc.) on the same Kubernetes Cluster as shown in Fig. 2. A Slurmd pod is made of two containers: Munged and Slurmd. They communicate through a UNIX socket on a shared directory. One of the main challenges in HPC is to reduce communications between HPC nodes. Slurmctld pods can be located anywhere in the Kubernetes infrastructure because they perform small communications with Slurmd pods during the bootstrap of the HPC job. Slurmd pods should gather on the same Cloud nodes to minimize communications across the physical network.

As a consequence, we will only consider the placement of pods. We used four physical nodes with 80 cores each on a private Cloud at USPN. Physical nodes are connected through a 1Gbps Ethernet network. As performance is not a concern for us, we used virtual machines as Cloud nodes to vary cores topology. We did not run any benchmark in pods. Unfortunately, Kubernetes default scheduler [21] dispatch pods horizontally in a round-robin way. The decision process is done as follows:

1. The scheduler enumerates registered and healthy nodes.
2. The scheduler runs the predicate tests to exclude unsuitable nodes. The remaining nodes form a group of potential nodes.
3. The scheduler runs priority tests against the potential nodes. Candidates are ordered by their score, with the highest ones on the top. At this point, the highest-scoring nodes are elected. They are put on a final list.
4. The Kubernetes default scheduler selects the winning node in a round-robin fashion to spread containers equally among the nodes.

This decision process is not very suitable for our use case because we prefer to have a distribution equivalent to the distribution supplied by the namespace scheduler described in Algorithm 1.

5.1   First Scenario: HPC Dedicated Kubernetes Cluster

This first round of experimentation will compare the Kubernetes default scheduler and our container scheduling strategy. We run 20 Slurmd pods in a private cloud with four nodes. Each node has 20 cores, and each pod requires 1 CPU.

Figure 3 illustrates the spreading of Slurmd pods on the Cloud nodes with the default scheduler. We can see that Slurmd pods are equally horizontally spread between Cloud nodes. Each Cloud node handles five pods.

Figure 4 illustrates the distribution of 20 Slurmd pods (grouped on a single namespace) on the Cloud nodes using our new container scheduling strategy. Slurmd pods run on a subset of 2 nodes instead of 4 previously. 17 Slurmd pods are gathered on Cloud node 2 and the 3 remaining pods on Cloud node 3. The node spreading is vertical. This distribution is more suited for Slurmd pods because using a physical network for communications induces a latency overhead.

Figure 5 illustrates the distribution of 20 Slurmd pods (grouped on two namespaces, i.e., two Slurm clusters) on the Cloud, nodes using our new container scheduling strategy. In this experience, the two namespaces are distributed as following: namespace cluster1 contains 9 Slurmd pods, and namespace cluster2 contains 11 pods. As a result, we can see that pods belonging to each Namespace are gathered together on different Cloud nodes.

From Figs. 4 and 5 we note that only two nodes are used to schedule 20 pods. However, with the default Kubernetes scheduling strategy, all nodes are used (4 nodes).

In Fig. 6, we expanded the experimentation to a larger scale. Our new test is made in a private cloud with four nodes, and each node has 80 cores. We created 280 pods randomly dispatched in 6 namespaces, i.e., 6 Slurm clusters. The division is 45 pods for Namespace $Cluster_1$, 41 pods for namespace $Cluster_2$, 42 pods for namespace $Cluster_3$, 62 pods for namespace $Cluster_4$, 46 pods for namespace $Cluster_5$, and 44 pods for namespace $Cluster_6$. Pods are randomly instantiated on our Kubernetes cluster. We can see a strong affinity between Cloud nodes and pods belonging to the same namespace.

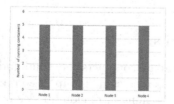

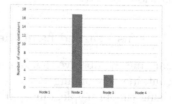

Fig. 3. Distribution of 20 pods between 4 nodes (20 cores each one) using the default kubernetes scheduling strategy

Fig. 4. Distribution of 20 pods (belongs to 1 namespace) between 4 nodes (20 cores each one) using our scheduling strategy

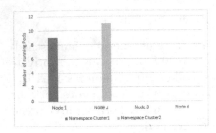

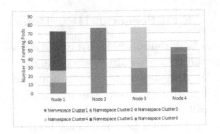

Fig. 5. Distribution of 20 pods (belongs to 2 namespaces) between 4 nodes (20 cores each one) using our scheduling strategy

Fig. 6. Distribution of 280 pods (belongs to 6 namespaces) between 4 nodes (80 cores each one) using our scheduling strategy

5.2   Second Scenario: HPC Cohabitation with Regular Pods

In this section, we present the cohabitation between our scheduling strategy and the default Kubernetes scheduling strategy. We created 6 Slurmd pods with 10 CPUs per pod in a dedicated namespace HPC and 240 regular pods with 1 CPU per pod in the default namespace to achieve this goal. We ran two experiments. For the first one, we used the default scheduler for every pod (Slurmd and regular). Then, we ran the same deployment for the second experimentation using our container scheduling strategy for Slurmd and regular pods.

The results of the first experiment are shown in Fig. 7. The Slurmd pods are spread on the four Cloud nodes in a round-robin way as expected. However, with our scheduling strategy, for Slurmd pods, we can see in Fig. 8 that all our Slurmd pods run on Cloud node 1. A Slurmd pod requires a lot of CPUs. When a pod runs on a Cloud node, it takes more CPUs than regular nodes.

The default scheduler spreads pods to other nodes to balance the overall load. When a new Slurmd pod arrives, our scheduling strategy tries to gather it with the others Slurmd pods. Then, the default scheduler tries to fill in free CPUs of other Cloud nodes with regular pods to aim for an overall balance of all Cloud nodes. Our Namespace scheduler integrates well with the default scheduler in a cohabitation context.

5.3   Lessons Learned from the Experiments

Summarizing, we observe that our scheduling strategy compacts the cloud nodes with containers belonging to the same namespaces (the same group of containers). Using this principle, our approach reduces the number of cloud nodes used and the cost of communication between containers. Our policy can cohabit with other container scheduling strategies easily. That means that the user can use our plan without changing its working environment. If he has already deployed his Kubernetes cluster, he can use the default Kubernetes scheduling strategy. Then it can also use our strategy on the same Kubernetes cluster to reduce the cost of communication between containers.

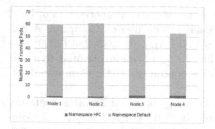

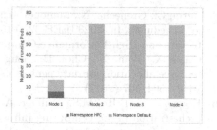

Fig. 7. Cohabitation between Slurmd and regular pods with default scheduler on 4 nodes (80 cores each one)

Fig. 8. Cohabitation between Slurmd pods with namespace scheduler and regular pods with default scheduler on 4 nodes (80 cores each one)

6   Conclusion

We have presented, in this paper, a new container scheduling strategy adapted for the HPC jobs scheduler. The novelty of our approach is to group the containers according to their Namespace in the same compute node to reduce the communication cost between containers.

Our strategy is validated in the context of the HPC ecosystem. As the first perspective, we propose to validate our approach in other areas as BigData and IA ecosystems.

In the Kubernetes framework, all cloud infrastructure nodes are used even if there are few running containers. This massive use of compute nodes increases the waste of resources. To reduce the waste of resources, we will propose as a second perspective a new consolidation approach that dynamically adapts the number of reserved cloud nodes depending on the total load on the nodes.

In our approach, we suppose that each container assignation is performed without any fault. As a third perspective, we propose to work on the fault-tolerance problem in case one node fails. The fault-tolerance problem can be investigated, for instance, by duplication of the containers, but the technique increases the system load. The choice of the number of duplicates is, in the context of many namespaces, one issue.

References

1. Bauer, M.: Solving Problems in HPC with Singularity. CernVM Workshop 2019, June 2019. https://cds.cern.ch/record/2677637
2. Beltre, A.M., Saha, P., Govindaraju, M., Younge, A., Grant, R.E.: Enabling HPC workloads on cloud infrastructure using kubernetes container orchestration mechanisms. In: 2019 IEEE/ACM International Workshop on Containers and New Orchestration Paradigms for Isolated Environments in HPC (CANOPIE-HPC), pp. 11–20 (2019)
3. Casalicchio, E., Iannucci, S.: The state-of-the-art in container technologies: application, orchestration and security. Concurrency Comput. Practice Exp. **32**(17), e5668 (2020)

4. Cérin, C., Greneche, N., Menouer, T.: Towards pervasive containerization of HPC job schedulers. In: 2020 IEEE 32nd International Symposium on Computer Architecture and High Performance Computing (SBAC-PAD), pp. 281–288 (2020)
5. Guan, X., Wan, X., Choi, B., Song, S., Zhu, J.: Application oriented dynamic resource allocation for data centers using docker containers. IEEE Commun. Lett. **21**(3), 504–507 (2017)
6. Hoque, S., d. Brito, M.S., Willner, A., Keil, O., Magedanz, T.: Towards container orchestration in fog computing infrastructures. In: 2017 IEEE 41st Annual Computer Software and Applications Conference (COMPSAC), vol. 2, pp. 294–299, July 2017
7. Jiang, C., et al.: Characterizing co-located workloads in alibaba cloud datacenters. IEEE Trans. Cloud Comput., 1 (2020)
8. Liu, B., Li, P., Lin, W., Shu, N., Li, Y., Chang, V.: A new container scheduling algorithm based on multi-objective optimization. Soft Comput. **22**, 1–12 (2018)
9. Marzolla, M., Babaoglu, Ö., Panzieri, F.: Server consolidation in clouds through gossiping. In: 12th IEEE International Symposium on a World of Wireless, Mobile and Multimedia Networks, WOWMOM, Lucca, Italy, 20–24 June, 2011, pp. 1–6 (2011)
10. Menouer, T., Darmon, P.: A new container scheduling algorithm based on multi-objective optimization. In: 27th Euromicro International Conference on Parallel, Distributed and Network-based Processing, Pavia, Italy, February 2019
11. Menouer, T., Manad, O., Cérin, C., Darmon, P.: Power efficiency containers scheduling approach based on machine learning technique for cloud computing environment. In: Esposito, C., Hong, J., Choo, K.-K.R. (eds.) I-SPAN 2019. CCIS, vol. 1080, pp. 193–206. Springer, Cham (2019). https://doi.org/10.1007/978-3-030-30143-9_16
12. Pahl, C., Lee, B.: Containers and clusters for edge cloud architectures - a technology review. In: 2015 3rd International Conference on Future Internet of Things and Cloud, pp. 379–386, August 2015
13. Smarr, L., Catlett, C.: Metacomputing. Commun. ACM **35**, 44–52 (1992)
14. Steve Buchanan, Janaka Rangama, N.B.: Introducing azure kubernetes service: a practical guide to container orchestration. In: Apress (2019)
15. Sureshkumar, M., Rajesh, P.: Optimizing the docker container usage based on load scheduling. In: 2017 2nd International Conference on Computing and Communications Technologies (ICCCT), pp. 165–168, February 2017
16. Xin, L.: The evolution of large-scale co-location technology at alibaba, 28 November 2019. https://www.alibabacloud.com/blog/the-evolution-of-large-scale-co-location-technology-at-alibaba_595595
17. Zhao, A., Huang, Q., Huang, Y., Zou, L., Chen, Z., Song, J.: Research on resource prediction model based on kubernetes container auto-scaling technology. IOP Conf. Ser. Materials Sci. Eng. **569**, 052092 (2019)
18. Zhou, N., Georgiou, Y., Zhong, L., Zhou, H., Pospieszny, M.: Container orchestration on HPC systems. In: 2020 IEEE 13th International Conference on Cloud Computing (CLOUD), pp. 34–36. IEEE (2020)
19. The apache software foundation. mesos, apache. http://mesos.apache.org/
20. Docker swarmkit. https://github.com/docker/swarmkit/
21. Kubernetes scheduler. https://kubernetes.io/

Architecture of an On-Time Data Transfer Framework in Cooperation with Scheduler System

Kohei Yamamoto[1]($\boxtimes$) , Arata Endo[2] , and Susumu Date[3]

[1] Graduate School of Information Science and Technology, Osaka University, Suita, Osaka 565-0871, Japan
yamamoto.kohei@ais.cmc.osaka-u.ac.jp
[2] Information Initiative Center, Nara Institute of Science and Technology, Ikoma, Nara 630-0192, Japan
[3] Cybermedia Center, Osaka University, Ibaraki, Osaka 567-0047, Japan

Abstract. Technological advancement in networking and IoT have given researchers new methods or techniques to perform numerical analysis and simulation with the latest data observed on IoT sensors and other measurement devices. In general, large-scale simulations necessitate high-performance computing (HPC) systems. Such HPC systems are operated in a shared manner among researchers. Therefore, it becomes inherently difficult for researchers to use the latest observation data generated on remote data sources for their simulations. To enable researchers to utilize fresh data on a remote data source for computation, we propose an on-time data transfer framework that enables the execution of jobs with fresh data generated on a remote site data on a shared HPC system by extending the SLURM scheduler. The proposed framework consists of two functions: *Job pinning* and *On-time data transfer*. With the job pinning function, the proposed framework prevents the scheduling algorithm from rearranging the scheduled start time of jobs. The on-time data transfer function is in charge of data transfer from a remote site to the data transfer node. It attempts to complete the data transfer at just the time of the pinned start time of jobs. The evaluation in this paper indicates that the proposed framework can keep data freshness high and minimize the job waiting time for data transfer.

Keywords: HPC · Staging · SLURM · Backfill algorithm · Data transfer

1 Introduction

Technological advancement in networking and IoT have given researchers new methods or techniques to perform numerical analysis and simulations with the

Supported by Japan Society for the Promotion of Science (JSPS) Grants-in-Aid for Scientific Research (KAKENHI).

latest data observed on IoT sensors and other measurement devices. An example
of such simulations is the data assimilation technique [9], which mixes observation data with input data and then processes them to obtain plausible results
statistically. Meteorologists use this technique to accurately predict the weather
after 30 min, using the latest 30-s observation data [13]. Also, in marine seis-
mological science, researchers have proposed a tsunami simulation method that
utilizes a data assimilation technique to obtain reliable results on tsunami height
and inundation volume before the arrival of the tsunami [16]. Considering the
popularization of IoT devices, it is easily predicted that more researchers will
be interested in applying such a technique to their simulations and numerical
analyses.

In general, large-scale simulations necessitate high-performance computing
(HPC) systems. Such HPC systems are operated in a shared manner among
researchers and the more the number of researchers, the greater the system
size. For this reason, most HPC systems have a resource management system,
a.k.a., a job management system or a scheduler deployed, so that each researcher
can benefit from high performance in a time-shared manner [10]. Therefore, it
becomes inherently difficult for researchers to use the latest observation data
generated on remote data sources for their simulations.

Today, researchers who need to utilize the "freshly" generated data on remote
data sources such as IoT devices for their large-scale analyses and simulations,
in most cases, use an HPC system designed for the dedicated use of a specific
research group. Furthermore, dedicated network links must also be prepared
to feed the data obtained or observed on measurement devices to the HPC
system. Taking the advancement of network technologies and the prevalence of
IoT devices into consideration, however, a framework that can feed the data
located outside the HPC system to the simulation and to numerical analysis
right after data generation will be essential in the future. In this paper, we
present a framework that we designed for this purpose.

This paper is structured as follows. In the next section, we describe the problem using fresh data on the shared HPC system and then provide the technical
requirements. In Sect. 3, we describe the design and implementation of our proposed framework. In Sect. 4, we investigate whether the proposed framework
meets the technical requirements through two experiments. In Sect. 5, we review
related works. Finally, Sect. 6 concludes this paper and discusses future works.

2   Research Goal and Technical Requirements

2.1   Research Goal

Under todays' HPC system design, there is an assumption that the data required
for program execution must be located in advance. In other words, when a user
submits a job, the user must place the data onto the HPC system before job submission. This fact means that the data prepared for program execution becomes
outdated under the current HPC system design, mainly when the HPC system
is heavily utilized.

To enable researchers to utilize the fresh data on a remote data source for their computations, the scheduler system, whose role is as the core resource management system on today's HPC system, is extended in this paper. Furthermore, the ultimate goal of this research is to re-consider what the scheduler system should be like, by overturning the current CPU-centric design of HPC systems, which assumes that data is located on HPC systems.

2.2 Approach for Feeding the Fresh Data on a Remote Data Source to the HPC System

Most scheduler systems available today offer a functionality that allows users to move data in the HPC system in cooperation with the job scheduler [12]. The data transfer is called *staging*. For example, staging allows a program to move data from a parallel filesystem as global storage to the local SSD of compute nodes before its execution and vice versa. However, this staging operation assumes the data itself is located inside the HPC system. Therefore, this staging operation cannot be applied to data transfer between the HPC system and a remote data source.

A straightforward way to perform the data transfer from a remote data source to the HPC system is to build a functionality that allows the job to retrieve the data directly from the data source on time just before job execution into the scheduler system. To realize such a functionality, when and how the data transfer is started becomes a key factor because it highly affects the relationship between data freshness and system utilization. Below are the three possible ways to transfer data.

The first way is the data transfer before job submission. This is usually used when the users submit jobs to the HPC system. As described in Sect. 1, the data transfer before a job submission makes the data necessary for program execution outdated.

The second way is to start the data transfer immediately after the commencement of job execution. The easiest way to utilize this data transfer is to put data transfer instructions such as *scp* and *wget* before program execution instruction such as *mpirun* [2] in the job script file submitted to the scheduler. This timing of data transfer enables user jobs to utilize the latest data generated on remote data sources. This second way seems a good strategy at a glance. However, the waiting time of data transfer completion from the corresponding job being dispatched causes the loss of CPU time, which results in the low job throughput of the HPC system.

The two ways above have a trade-off problem between job throughput and data freshness. For this reason, this research explores a possible third way as an intermediate way between the above two ways; namely, data transfer based on the job's scheduled start time. We refer to this way as 'predictive data transfer' in this paper. This data transfer way completes the data transfer just at the time the job starts. In other words, the temporal gap between data transfer completion and job start is minimized, so that system utilization is not lowered. Moreover, this predictive data transfer aims to achieve both job throughput and data freshness. However, transferring data from a data source to the local HPC system just before job execution starts is challenging in terms of how the completion of data transfer is done on time before job execution.

2.3 Technical Requirements

To build predictive data transfer between an HPC system and a remote data source, we use a strategy that manages both job scheduling and data transfer scheduling. For this purpose, we extend SLURM [17] as its job scheduler and the backfill algorithm [15] as its job scheduling algorithm. SLURM is a resource management system provided as open-source software designed to be pluggable depending on the administration policy. The backfill algorithm is a scheduling algorithm that creates a resource allocation map on which user-submitted jobs are allocated to computational resources over time. The resource allocation map is rearranged flexibly to the actual status of job execution on the HPC system. The combination of SLURM and this algorithm both have a rich track record of use in HPC systems.

In this research, we have set the following three technical requirements to design the framework that allows users to retrieve remote data: *High data freshness*, *Minimum job waiting time for data transfer* and *Selectivity*.

High Data Freshness. As illustrated in Fig. 1, we define *data freshness* as the time from the start of data transfer to the beginning of job execution. We assume that the latest data is transferred from the remote data source to the HPC system. High data freshness means that user jobs can use the latest data generated on a remote data source.

Minimum Job Waiting Time for Data Transfer. As described, it is ideal that a job can use the data just generated on a remote source when it starts. If the job has to wait for the data arrival after it is dispatched to CPU resources, system utilization is lowered. The waiting time for data transfer after job execution starts should be minimized not to lower the job throughput on the HPC system.

Selectivity. Some users may want to perform computation using the fresh data even if the job start is delayed; others may not have time constraints. Suggested features should require some means of allowing users to select whether the user will use them or not.

3 Proposal

3.1 Overview

In this paper, we propose an on-time data transfer framework to enable the execution of jobs with fresh data generated on a remote site on the shared HPC

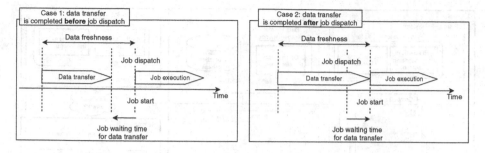

Fig. 1. Description of data freshness and job waiting time for data transfer.

system. This framework is designed to be deployed in a wide-area computing environment. The framework is configured onto the local site where an HPC system is located and remote sites where data are generated. Figure 2 shows the overview of the proposed framework. The framework is characterized by a data transfer module extended on the top of the scheduler system. The data transfer module spans both local and remote sites. It is designed to interact with the scheduler on the HPC system and manage data transfer between the local site and remote data sources.

The proposed framework consists of two functions: *Job pinning* and *On-time data transfer*. The former function is deployed in the job management node where the job scheduler server runs. The latter function is located onto the data transfer management node and data transfer nodes. The basic idea behind this composition is that we have both data transfer and job scheduling managed in the proposed framework. For this management, job scheduling information such as scheduled start times of jobs and data transfer status are shared in the framework. For the implementation, SLURM [17] was used for the job pinning function. FastAPI [1], a server framework in Python, was used for the on-time data transfer function.

The proposed framework works as follows. With the job pinning function, the proposed framework prevents the scheduling algorithm from rearranging the scheduled start time of jobs. The function informs the data transfer management node of the pinned scheduled times of jobs. The on-time data transfer function is in charge of data transfer from a remote site to the data transfer node. It attempts to complete the data transfer just at the time of the pinned start time of jobs.

3.2   The Job Pinning Function

Under the HPC system with the backfill algorithm, the resource map is often changed to raise the job throughput. If some jobs finish earlier than expected, the algorithm attempts to move the jobs scheduled later than the finished jobs. This algorithm is effective for improving the job throughput and shortening waiting time. However, when we consider how we feed the fresh data generated on a remote data source to the programs executed on the HPC system, this

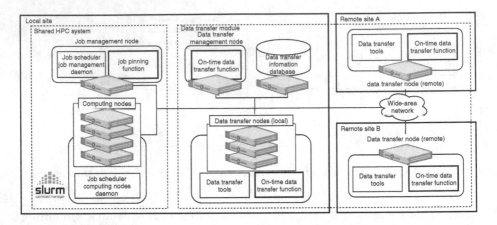

Fig. 2. The architecture of the proposed framework.

flexibility and adjustability brought by this algorithm make the feed of the fresh data that arrived just before job execution much more difficult.

From the consideration above, we have introduced a job pinning function that instructs the backfill algorithm to pin the start time of jobs once the algorithm allocates them. In short, this function does not allow the scheduler to change the job scheduled start time. To use this function, the users need to add the four lines shown in Table 1 to the job script. If the user specifies the use of the pinning function, this pinning function notifies the on-time data transfer function of the information necessary for data transfer. The information includes the pinned scheduled start time, job ID, source file path, destination file path, and transfer data size.

3.3   The On-Time Data Transfer Function

This function has the responsibility of completing the data transfer just by the pinned scheduled start time. Furthermore, to prevent a job from starting before the arrival of data necessary for its job, this function monitors the progress of the data transfer. The data transfer management node determines when to start the data transfer. Currently, only the maximum transfer bandwidth available is

Table 1. The settings to be added to the job script and their contents.

Name	Content
#SBATCH --pin	Declare the use of a function
#SBATCH --srcfile=	Source file path
#SBATCH --dstfile=	Destination file path
#SBATCH --srcfilesize=	Transfer file size

considered although there is room for discussion on when the algorithm starts the data transfer. The Job ID and a pair of IP addresses are used for the management of data transfer.

When the job pinning function pins the job scheduled start time, the job management node notifies the pinned time to the data management node. Upon receiving a notification, this function performs on-time data transfer in the following three steps: data transfer scheduling, data transfer executing, and inquiry-responding. In the first data transfer scheduling step, the on-time data transfer function determines the data transfer start time according to the data size and available network bandwidth. In the second data transfer executing step, the on-time data transfer function causes the data transfer node to perform the data transfer. The data transfer management node manages when the data transfer started and ended in the data transfer database. In the third inquiry-responding step, the on-time data transfer function responds to inquiries from the job scheduling algorithm about whether the data transfer node has completed the data transfer or not. If the data transfer has been completed, the job pinning function allows the job scheduler to assign the job to the compute nodes.

4   Evaluation

To investigate the effectiveness of the proposed framework, we perform two experiments to confirm whether the proposed framework functions correctly and we investigate how the proposed framework affects job throughput. In the first experiment, we observe that the proposed framework can execute jobs with high data freshness while keeping the job waiting time for data transfer short. In the second experiment, we evaluate the incurred overhead of the job throughput by the job pinning function in the proposed framework.

4.1   Experimental Procedure

We submit a set of 20 jobs with two-second intervals between jobs. This submission test is repeated ten times. Each job executes the *sleep* command as a dummy job for a randomly pre-specified time from 5 to 20 min, considering there exist different jobs in HPC systems. Each job is set to request one to three computing nodes. These jobs are executed in parallel. To confirm that the proposed framework functions correctly, we observe and compare the behavior of the three methods described in Sect. 2.2 in terms of data freshness and job waiting time with the data-transfer size changed. The three methods are the proposal, the *before-submission stored* (BSS) method, and the *after-dispatch stored* (ADS) method. In this experiment, among the 20 job sets, the number of jobs that feed the latest data is one. For investigation of the overhead, the ratio of such jobs is changed to observe how the ratio affects job throughput. We measure the time from the submission of the first job to the end of all jobs.

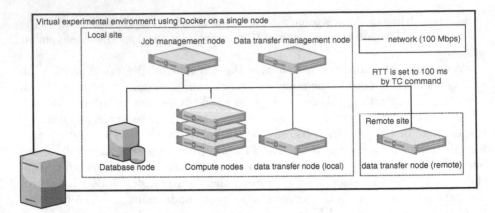

Fig. 3. The experimental environment.

Table 2. Software used for experiment and the spec of an experiment node.

Software	Version
SLURM	19.05.5
Python	3.6.8
Fast API	0.1.0
Docker	20.10.2
Name	**Spec**
CPU	Intel(R) Xeon(R) Gold 5218 CPU @ 2.30 GHz
Memory	DDR4-2933 128 GB
OS	Ubuntu 20.04 LTS

4.2 Experimental Environment

We have built the simulated experiment environment shown in Fig. 3 using Docker [4] on a single experiment node. The HPC system is comprised of three computing nodes, a database node, a job management node, a local data transfer node, a remote data transfer node, and a data management node. In the job management node, the duration of the resource map was set to 24 h. The software and the specification of the nodes used for this experiment are shown in Table 2. The RTT (Round-Trip Time) between the HPC system and remote site was set at 100 ms to experiment on a simulated wide-area network environment. All the network bandwidths were set to 100 Mbps. We used the Traffic Control (TC) command [3] for the configuration and setting of latency. We assume that the computing nodes mount the file system of the local data transfer node.

4.3 Effectiveness of the Proposed Framework

The graph shown in Fig. 4 plots the result of data freshness and the job waiting time from the time when the job is dispatched to when its execution starts. In the graph, the blue, green, and red dots indicate that the data size is 1, 5, and 10 GB, respectively. Note that the job waiting time for the data transfer takes a negative value if the data transfer finishes before the job is dispatched. The plot is the average value of all 20 jobs using the fresh data submitted for the experiment. Apparently, the ADS increases the job waiting time from when it is dispatched to when its execution started in proportion to data size. In BSS, the negative value of the job waiting time is much larger than the proposal because the job is submitted after the data transfer is completed. These results are reasonably understood. A remarkable result to be mentioned from this graph is that the proposed framework is superior to BSS in data freshness for each data size. This result means that the proposed framework has achieved retrieving the data generated on a remote site to the HPC system while keeping the job waiting time short.

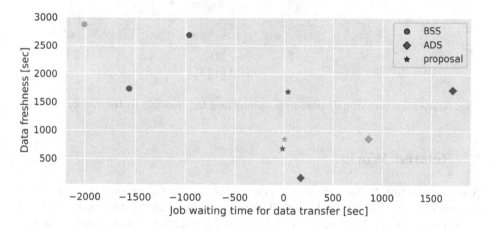

Fig. 4. Average data freshness and job waiting time. (Color figure online)

4.4 Overhead of the Job Scheduling Algorithm

Figure 5 shows the average turnaround time from when the first job is submitted to when the last job is finished. The "ratio" is the ratio of jobs that requires fresh data on the remote source to jobs that do not need it. For example, a ratio of zero percent means that the set of 20 jobs does not contain any jobs that utilize the proposed framework. Also, a ratio of five percent means that one job in 20 jobs utilizes the proposed framework.

We observed that the turnaround time was almost the same in the case of the ratio of zero, five, and ten percent. On the other hand, when the ratio was 15% and 20%, we observed that the turnaround time increased by about 1500 s compared to the ratio of zero. This fact indicates that the proposed framework can maintain job throughput when the ratio is small. Still, when the ratio is more than 15%, the overhead of the job scheduling algorithm becomes larger, and the job throughput decreases.

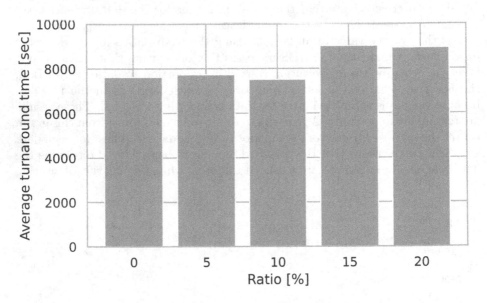

Fig. 5. Average turnaround time from submitting the first job to the end of all jobs.

5   Related Works

A timely staging framework to integrate wide-area data transfer with an HPC system has been presented in [14]. The main focus of this framework [14] is to minimize the period where data to be used for computation is located on the HPC system. This research [14] shows how the completion of input data staging can be coincident with the job start-up, but does not put emphasis on data freshness. Our research focuses on feeding the fresh data on a remote site to the HPC system.

Data transfer frameworks in a wide-area network have been presented in [5,11,18]. DTN-as-a-Service (DaaS) framework [18] provides high-performance network data transfers using an integration of techniques, including virtualization, network provisioning, and performance data analysis. BigData Express [11] provides a schedulable, predictable, and high-performance data transfer service to address the high-performance and time-constraint challenges of data transfer required by extreme-scale science applications. The Globus Striped GridFTP Framework [5], which is a set of client and server libraries designed to support the construction of data-intensive tools and applications using GridFTP [6], performs the data transfer with reliability and high performance by extending the

FTP. These data transfer frameworks focus on making the data transfer more efficient and effective and can be used as a data transfer method in our research.

For the data movement inside the HPC system, several researches have been presented [7,8,12]. In NORNS [12], an asynchronous staging service has been proposed to handle data dependencies between the different phases in users processing workflows by extending SLURM [17]. A data elevator [7] intercepts the I/O calls and then asynchronously transfers the data to the final destination in the background, targeting HPC systems that support a multi-level file system. These studies perform staging to improve I/O performance in the HPC system. A dynamic traffic control method [8] focuses on the traffic collision of the inter-node communication and the staging traffic to improve job throughput, targeting the cluster system. These authors do not assume that the input data outside the HPC system is directly transferred to the computing nodes.

6   Conclusion

This paper has proposed an on-time data transfer framework by extending SLURM to enable the execution of jobs with fresh data generated on a remote site. Our proposed framework pins the start time of such jobs for on-time data transfer just by the execution start. The evaluation in this paper indicates that the proposed framework can achieve retrieval of the data generated on a remote site to the HPC system while keeping job waiting time short.

However, we must tackle a few issues to improve our proposed framework in the future. The first issue is to improve the accuracy of the data transfer scheduling function. The proposed framework determines the transfer start time based on the data size and the network bandwidth measured in advance. However, available network bandwidth changes over time. To improve the proposed framework to be more practical, we need to design and adopt a sophisticated algorithm for data transfer that considers the dynamicity of network traffic. Another issue is to control multiple data transfer requests simultaneously. We assume that the ratio of jobs that feed the fresh data is not very large and that the data transfer using the proposed framework does not co-occur. However, considering the recent prevalence of IoT devices and the development of networks, the number of such jobs will increase. It is necessary to support the control of multiple data transfer requests simultaneously.

Acknowledgements. This research is partly supported by JSPS KAKENHI Grant Number JP17KT0083 and JP21K11912.

References

1. FastAPI. https://github.com/tiangolo/fastapi
2. MPI: A Message-Passing Interface Standard. https://www.mpi-forum.org/docs/mpi-3.0/mpi30-report.pdf
3. TC(8) - Linux Manual Page. https://man7.org/linux/man-pages/man8/tc.8.html

4. Why Docker? — Docker. https://www.docker.com/why-docker
5. Allcock, W., Bresnahan, J., Kettimuthu, R., Link, M.: The Globus striped GridFTP framework and server. In: Proceedings of the 2005 ACM/IEEE Conference on Supercomputing, p. 54, November 2005
6. Allcock, W.: GridFTP: Protocol Extensions to FTP for the Grid. http://www.ggf.org/documents/GFD.20.pdf (2003)
7. Dong, B., Byna, S., Wu, K., Prabhat, Johansen, H., Johnson, J.N., Keen, N.: Data elevator: low-contention data movement in hierarchical storage system. In: Proceedings of the 2016 IEEE International Conference on High Performance Computing (HiPC), pp. 152–161, December 2016
8. Endo, A., et al.: Dynamic traffic control of staging traffic on the interconnect of the HPC cluster system. IEEE Access **8**, 198518–198531 (2020)
9. Ghil, M., Malanotte-Rizzoli, P.: Data assimilation in meteorology and oceanography. Adv. Geophys. **33**, 141–266 (1991)
10. Hovestadt, M., Kao, O., Keller, A., Streit, A.: Scheduling in HPC resource management systems: queuing vs. planning. In: Proceedings of the 2003 Job Scheduling Strategies for Parallel Processing (JSSPP), pp. 1–20, June 2003
11. Lu, Q., et al.: BigData express: toward schedulable, predictable, and high-performance data transfer. In: Proceedings of the 2018 IEEE/ACM Innovating the Network for Data-Intensive Science (INDIS), pp. 75–84, November 2018
12. Miranda, A., Jackson, W., Tocci, T., Panourgias, I., Nou, R.: NORNS: extending Slurm to support data-driven workflows through asynchronous data staging. In: Proceedings of the 2019 IEEE International Conference on Cluster Computing (CLUSTER), pp. 1–12, November 2019
13. Miyoshi, T., et al.: Big data assimilation toward post-petascale severe weather prediction: an overview and progress. Proc. IEEE **104**(11), 2155–2179 (2016)
14. Monti, H.M., Butt, A.R., Vazhkudai, S.S.: On timely staging of HPC job input data. IEEE Trans. Parallel Distrib. Syst. **24**(9), 1841–1851 (2013)
15. Shmueli, E., Feitelson, D.G.: Backfilling with lookahead to optimize the performance of parallel job scheduling. In: Proceedings of the 2003 Job Scheduling Strategies for Parallel Processing (JSSPP), pp. 228–251, June 2003
16. Wang, Y., Satake, K., Maeda, T., Gusman, A.R.: Data assimilation with dispersive tsunami model: a test for the Nankai trough. Earth, Planets Space **70**(1), 1–9 (2018)
17. Yoo, A.B., Jette, M.A., Grondona, M.: SLURM: simple linux utility for resource management. In: Proceedings of the 2003 Job Scheduling Strategies for Parallel Processing (JSSPP), pp. 44–60, June 2003
18. Yu, S., et al.: SCinet DTN-as-a-service framework. In: Proceedings of the 2019 IEEE/ACM Innovating the Network for Data-Intensive Science (INDIS), pp. 1–8, November 2019

Storage

Data Delta Based Hybrid Writes for Erasure-Coded Storage Systems

Qiang Huang[1], Hui Chen[1(✉)], Bing Wei[1(✉)], Jigang Wu[1], and Limin Xiao[2]

[1] School of Computer Science and Technology, Guangdong University of Technology, Guangzhou, China
chenhui02@gdut.edu.cn, weibing@buaa.edu.cn
[2] School of Computer Science and Engineering, Beihang University, Beijing, China

Abstract. Erasure coding is widely used in storage systems since it can offer higher reliability at lower redundancy than data replication. However, erasure-coded storage systems have to perform a *partial write* to an entire erasure coding group for a small write, which causes a time-consuming write-after-read. This paper presents an efficient data delta based hybrid write scheme, named DABRI, which supports fast partial writes for erasure-coded storage systems. DABRI uses data deltas that are the differences between latest data values and original data values to bypass the computation of parity deltas and the read of old data. For a series of n partial writes to the same data, DABRI performs log-based data and parity updates for the first write, and takes in-place data updates and log-based parity updates for the last $n-1$ writes. This enables new data to be written into data nodes and parity nodes in parallel, and the overhead of data reads and parity updates can be mitigated. Based on DABRI, we design and implement an erasure-coded prototype storage system that can deliver high-performance for small-write-intensive applications. Experimental results on real-world traces show that DABRI can successfully improve the I/O throughput by up to 77.41%, compared with the state-of-the-art cited in this paper.

Keywords: Erasure coding · Update writes · Storage systems · Data deltas · Hybrid writes

1 Introduction

Component failures are the norm rather than the exception in large-scale storage systems [6,13,15]. For example, the average annualized failure rate for the 6 HDD make/model-based disk groups in large-scale cluster storage systems (with 100,000+ disks) is greater than 3% [8]. Replication and erasure coding (EC) are the two promising technologies for maintaining data durability against failures [19]. In replication, the file content is split into chunks and each chunk of the file is independently replicated at multiple disks or machines [18]. Three-way replication (R3) is widely used in modern storage systems [3]. In EC, the original

© IFIP International Federation for Information Processing 2022
Published by Springer Nature Switzerland AG 2022
C. Cérin et al. (Eds.): NPC 2021, LNCS 13152, pp. 171–182, 2022.
https://doi.org/10.1007/978-3-030-93571-9_14

data chunks are encoded to generate the parity chunks, such that the lost data chunks can be recovered by a subset of data and parity chunks [18]. It is known that EC introduces less storage overhead than replication while providing the same or even higher level of durability [4,9]. Therefore, many enterprise storage systems employ EC to ensure data reliability [2,5].

Two important factors affecting storage system performance are storage overhead and I/O overhead. Storage overhead refers to the amount of hard disk space consumed to store user data and ensure its robustness. I/O overhead refers to the average number of times a user initiates a request to read or write the hard disk. Although EC can significantly reduce storage overhead, it can introduce large I/O overhead under small update write scenarios [6,15]. This is because a small update write issues a partial write to an entire EC group, then the parity must be updated with the data, which causes extra computation and I/O overhead [14,16]. If the parity cannot be updated with the new data, then the new data cannot be constructed by the parity. Our work is to guarantee the consistency between data and parity. Real-world workloads of enterprise storage systems are often dominated by random small writes [1,17]. Therefore, partial writes are issued frequently.

In erasure-coded storage systems, there are two ways of parity updates: 1) reconstructed-write (RCW); and 2) read-modify-write (RMW) [1]. In RCW, all unmodified data of an EC group are read first, then they are encoded with the new data to compute the new parities. In RMW, the old data to be updated are read first, then the differences between the new data and the old data are calculated, finally the parity deltas are computed using the data difference. RCW usually outperforms RMW in large sequential write scenarios, whereas RMW usually outperforms RCW in small write scenarios [1]. Therefore, RMW is more suitable for partial writes.

Modern storage systems usually combine RMW and logging to speed up partial writes [11]. Full-logging (FL) saves the disk read overhead of parity chunks by appending all data and parity updates [5]. That is, after the modified data range and parity deltas are respectively sent to the corresponding data and parity nodes, the storage nodes create logs to store the updates. Parity-logging (PL) is proposed to reduce the update overhead of parity [7]. It updates data with in-place manner and uses logging to update parities, thereby reducing the I/O overhead of parity updates without affecting data reads. Parity-logging with reserved space (PLR) is built on PL, it takes a hybrid of in-place data updates and log-based updates to balance the I/O overhead of updates and recovery [1]. In addition, PLR reserves a certain amount of disk space next to each parity chunk such that the parity updates can be performed in the reserved space to mitigate disk seeks. However, these approaches still have to perform a time-consuming write-after-read for a partial write.

In this paper, we focus on how to reduce the execution time of partial writes for erasure-coded storage systems. We propose an efficient data delta based hybrid write scheme named DABRI to solve the problem. DABRI uses data deltas instead of parity deltas to bypass the computation of parity deltas and the read of old data. For a series of n partial writes to the same data, DABRI performs log-based data and parity updates for the first write. This enables data nodes to know the original data are needed by retrieving their own logs. As a result, I/O operations between data nodes and parity nodes can be overlapped. For the last $n - 1$ writes, the data nodes know the original data are not needed by retrieving their own logs. This enables the writes of new data between data nodes and parity nodes to be performed in parallel. Meanwhile, DABRI takes a hybrid of in-place data updates and log-based parity updates to balance the costs of partial writes and data reads. The main contributions of this paper are summarized as follows.

- We propose an efficient data delta based hybrid write scheme DABRI to support fast partial writes for erasure-coded storage systems. DABRI uses data deltas instead of parity deltas to bypass the computation of parity deltas and the read of old data. In DABRI, every data node has its own logs for appending new data, which enables data nodes to know whether original data are needed rather than receive responses from parity nodes. Therefore, data nodes can perform I/O operations before receiving responses from parity nodes. For a series of n partial writes to the same data, DABRI performs a write-after-read for the first write, and only a write for the last $n - 1$ writes.
- Based on DABRI, we design a distributed prototype file system for small-write-intensive workloads. We have compared DABRI with the latest work on the proposed storage system through the same real-world I/O trace used in [1]. Experimental results show that DABRI can successfully improve the I/O throughput by up to 77.41%.

2   Preliminary

We split file content into chunks and apply EC independently on a per-chunk basis. Let $EC(k, m)$ denote an EC scheme, then the $k + m$ stripes are called an EC group, in which k is the number of data chunks and m is the number of parity chunks. The m parity chunks are generated by encoding the k data chunks. We consider maximum distance separable EC, i.e., the original data chunks can be reconstructed from any k of the $k + m$ data and parity chunks for an EC group. In an EC group, each parity chunk can be encoded by computing a linear combination of k data chunks. For $EC(k, m)$, let $d_j (1 \leq j \leq k)$ denote a data chunk, $p_i (1 \leq i \leq m)$ denote a parity chunk, then p_i can be computed by

$$p_i = \gamma_{i1}d_1 + \gamma_{i2}d_2 + \cdots + \gamma_{ik}d_k \tag{1}$$

where $\gamma_{ij} (1 \leq j \leq k, 1 \leq i \leq m)$ denotes an encoding coefficient. All arithmetic operations are performed in the Galois Field $GF(2^w)$ [10].

In an EC group, the associated parity chunks must be updated when a data chunk is updated. Intuitively, the storage system can perform re-encoding, that is, use RCW to update the parity chunks. However, RCW must read all unmodified data of an EC group, which introduces high I/O overhead in small writes.

An alternative to compute new parity chunks is RMW, which takes advantage of the linearity property of EC to reduce the read amount of data. Assume data chunk d_l $(1 \leq l \leq k)$ is updated to d_l' in an EC group, then each parity chunk in the group must be updated. The new parity chunk p_i' $(1 \leq i \leq m)$ can be computed by

$$
\begin{aligned}
p_i' &= \sum_{j=1, j \neq l}^{k} \gamma_{ij} d_j + \gamma_{il} d_l' \\
&= \sum_{j=1}^{k} \gamma_{ij} d_j - \gamma_{il} d_l + \gamma_{il} d_l' \\
&= p_i + \gamma_{il}(d_l' - d_l) \\
&= p_i + \gamma_{il} \times \Delta d_l \\
&= p_i + \Delta p_i
\end{aligned}
\tag{2}
$$

where Δd_l is the data delta, Δp_i is the parity delta. According to Eq. (2), the new parity chunk is computed by the change of data, instead of re-encoding with the manner of summing over all the data chunks. When any part of a data chunk is updated, even for a single word, the corresponding parity delta can be computed by Eq. (2).

EC based storage systems usually combine RMW and logging to improve the performance of partial writes. Logging can significantly reduce the number of disk seeks for writes. However, log-based approaches introduce additional disk seeks to the update log for reads, because the data are scattered in the log. This in particular hurts the performance of sequential reads. PL updates data with in-place manner and uses logging to update parities. It can effectively improve the write performance without affecting data reads.

Let $d_l^{(r)}$ denote the rth write on a data d_l. According to Eq. (2), $d_l^{(r-1)}$ must be read from the data node before the new value $d_l^{(r)}$ is written into the data node. Then the new parity delta $\Delta p_i^{(r)} = \gamma_{il}(d_l^{(r)} - d_l^{(r-1)}) = \gamma_{il} \times \Delta d_l^{(r)}$ is computed and then written into parity nodes. The entire procedure of PL is illustrated in Fig. 1.

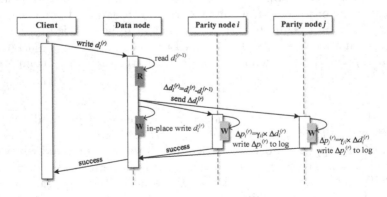

Fig. 1. Procedure of PL.

3 Data Delta Based Hybrid Writes

3.1 Computing Parities

In RMW, data nodes must read old data to compute parity deltas, which introduces a time-consuming write-after-read for a partial write. If the number of data reads can be reduced, then the execution time of partial writes can be reduced. Therefore, our goal is to reduce the number of data reads caused by partial writes.

Let $p_i^{(r)}$ denote the rth write on parity p_i, and p_i is corresponding to the data d_l in an EC group. Let $d_l^{(0)}$ and $p_i^{(0)}$ denote the original data of d_l and the original parity of p_i, respectively. Assume that d_l is overwritten r times, then we have $d_l^{(1)}, d_l^{(2)}, \cdots, d_l^{(r)}, p_i^{(1)}, p_i^{(2)}, \cdots, p_i^{(r)}$. According to Eq. (2), we have

$$
\begin{cases}
p_i^{(1)} = p_i^{(0)} + \Delta p_i^{(1)} = p_i^{(0)} + \gamma_{il}(d_l^{(1)} - d_l^{(0)}) \\
p_i^{(2)} = p_i^{(1)} + \Delta p_i^{(2)} = p_i^{(1)} + \gamma_{il}(d_l^{(2)} - d_l^{(1)}) \\
\vdots \\
p_i^{(r)} = p_i^{(r-1)} + \Delta p_i^{(r)} = p_i^{(r-1)} + \gamma_{il}(d_l^{(r)} - d_l^{(r-1)})
\end{cases}
\tag{3}
$$

according to Eq. (3), we have

$$
\begin{aligned}
p_i^{(r)} &= p_i^{(0)} + \Delta p_i^{(1)} + \Delta p_i^{(1)} + \cdots + \Delta p_i^{(r-1)} + \Delta p_i^{(r)} \\
&= p_i^{(0)} - \gamma_{il}d_l^{(0)} + \gamma_{il}d_l^{(1)} - \gamma_{il}d_l^{(1)} + \gamma_{il}d_l^{(2)} \\
&\quad - \cdots - \gamma_{il}d_l^{(r-2)} + \gamma_{il}d_l^{(r-1)} - \gamma_{il}d_l^{(r-1)} \\
&\quad + \gamma_{il}d_l^{(r)} \\
&= p_i^{(0)} + \gamma_{il}(d_l^{(r)} - d_l^{(0)})
\end{aligned}
\tag{4}
$$

Equation (4) illustrates that $p_i^{(r)}$ can be computed by $p_i^{(0)}$, $d_l^{(r)}$, and $d_l^{(0)}$. This enables us not to compute parity delta for each partial write, but compute data delta using $d_l^{(r)}$ and $d_l^{(0)}$. The number of reads can be significantly reduced by using data deltas instead of parity deltas. This is because $d_l^{(0)}$ that has been forwarded to parity nodes during the writing of $d_l^{(1)}$ can be reused for the writing of $d_l^{(r)}$ ($r \geq 2$). As a result, the read of $d_l^{(r-1)}$ can be bypassed when $r \geq 2$.

According to Eq. (4), the data node must send $d_l^{(0)}$ to parity nodes. However, it does not know whether $d_l^{(0)}$ is needed only if receiving the responses from parity nodes. It is costly to maintain the consensus about whether $d_l^{(0)}$ is needed for every chunk on parity nodes. This is because maintaining the consensus can cause large communication overhead, memory footprint, and design complexity. Therefore, we propose DABRI to improve these limitations.

3.2   Proposed DABRI

Based on the analysis, we design a data delta based hybrid write scheme, named DABRI, to reduce the number of data reads caused by partial writes. Figure 2 shows the complete partial write procedure of DABRI. For each partial write, the client first forwards the new data $d_l^{(r)}$ to the data node, then the data node forwards $d_l^{(r)}$ to the parity nodes. The original data value $d_l^{(0)}$ is read in the 1st partial write, whereas it will no longer be read in subsequent partial writes.

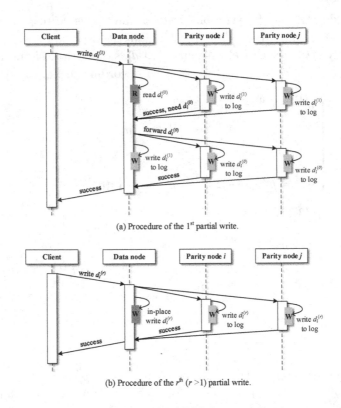

(a) Procedure of the 1st partial write.

(b) Procedure of the rth ($r > 1$) partial write.

Fig. 2. Procedure of DABRI for partial writes.

Figure 2(a) shows the procedure of the 1st partial write. When receiving $d_l^{(1)}$, the data node knows that $d_l^{(0)}$ has not been updated because no updated data can be retrieved in its log. Then the data node knows that the parity nodes need $d_l^{(0)}$, hence it reads $d_l^{(0)}$ before receiving the explicit requests sent from the parity nodes. When receiving $d_l^{(1)}$, the parity nodes append $d_l^{(1)}$ to their own logs, then explicitly request $d_l^{(0)}$ asking the data node. Once receiving the request, the data node immediately sends $d_l^{(0)}$ to the parity nodes. The reading of $d_l^{(0)}$ on the data node and the writing of $d_l^{(1)}$ on the parity nodes can be performed in parallel.

The data node appends $d_l^{(1)}$ to its log after sending $d_l^{(0)}$. Once receiving $d_l^{(0)}$, the parity nodes append it to their own logs and then return success to the parity node. The writing of $d_l^{(1)}$ on the data node and the writing of $d_l^{(0)}$ on the parity nodes can also be performed in parallel.

Figure 2(b) shows the procedure of the rth ($r > 1$) partial write. When receiving $d_l^{(r)}$, the data node knows that $d_l^{(0)}$ has been updated because the $d_l^{(1)}$ can be retrieved in its log. Then the data node in-place writes $d_l^{(r)}$ into the original file, and it will not read and forward $d_l^{(0)}$ again. When receiving $d_l^{(r)}$, the parity nodes append $d_l^{(r)}$ to their own logs without requesting $d_l^{(0)}$. The writing of $d_l^{(r)}$ between the data node and the parity nodes can be performed in parallel.

3.3 Merging Compactions

As write operations execute, the size of the logs located on data nodes and parity nodes increases largely. When the utilization of a node (ratio of used disk space at the node to total capacity of the node) reaches a threshold, merging compactions are performed asynchronously. This compaction process brings the benefits as follows: 1) the disk usage of the logs can be largely shrunk; and 2) the amount of data that must be read from the logs can be largely reduced, especially during the recovery of lost data. Merging compactions can also be triggered manually by the system administrators. A compaction process traverses all logs in an EC group. For data chunks, we read the latest data on the log and then write them into the original file with in-place manner. After that, we invalidate the data that have been read on the log. For parity chunks, we use the original data $d_l^{(0)}$ and the latest data $d_l^{(r)}$ to update $p_i^{(0)}$, then invalidate $d_l^{(0)}$ and $d_l^{(r)}$. If a log does not contain any valid data, then it can be deleted.

4 Implementation

Based on DABRI, we implement a prototype of distributed file system named HWDFS. HWDFS splits file content into fixed-size data chunks, it stores each chunk at a single data node. HWDFS encodes each k consecutive data chunks of a file to generate m parity chunks. The size of a parity chunk is the same as that of a data chunk, each parity chunk is independently stored on a single parity node. Similar to GFS [12] and HDFS [3], HWDFS implements a global master (metadata node) to maintain all file system metadata. The master chooses the node to host a data chunk or a parity chunk. When reading a file, the HWDFS client first asks the master for the location information of the chunks of the file. It then contacts the data node that holds the target chunk for data transfer. When writing a file, the HWDFS client first asks the master to choose the nodes to host the data chunk and the corresponding parity chunks.

Figure 3 shows the procedure of HWDFS for partial writes. The client first asks the master for metadata, then it forwards the write request to the corresponding data node. When receiving the write request, the data node directly

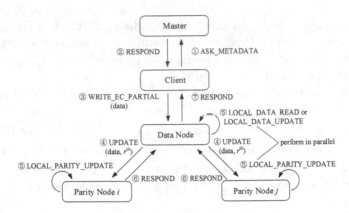

Fig. 3. Procedure of HWDFS for partial writes.

forwards it to all relevant parity nodes (parity nodes i and j). After that, the data node immediately performs local data read (when $r = 1$) or local data update. When receiving the write request from the data node, the parity nodes perform local parity update to the per-chunk log. The local parity update (step 5) of the parity nodes can be performed in parallel with the local data read or local data update (step 5) of the data node.

5   Evaluation

Our experiments are conducted on 8-node machines, these machines are homogeneous (heterogeneous machines are also acceptable), four of which are the data nodes, two of which are the parity nodes, one of which is the client, and the last one is the master. Each machine is configured with two 20-core 2.2 GHz Intel Xeon 4114 CPUs, 128 GB of memory, four 4 TB disks, and an Ubuntu 18.04 LTS operating system. The network is 10-Gigabit Ethernet. The size of each data or parity chunk is 64 MB. For an $EC(k, m)$ group, k and m are set to 4 and 2, respectively. We evaluate the proposed DABRI by comparing with the following three approaches: 1) PL [7]; 2) PLR [1]; and 3) R3 [3]. In R3, the file content is split into chunks, each chunk of the file is independently replicated at three data nodes, data are updated with in-place writes. Each experiment is repeated five times and the mean values are shown. In our evaluation, PL and R3 are implemented into HWDFS, PLR is implemented using the source code provided in [1]. In PLR, the reserved space size for every parity chunk is equal to the chunk size. The source code of HWDFS is available for public-domain use on https://github.com/MrBrutalchiefs/HWDFS.

5.1   Trace Evaluation

This section evaluates the performance of all approaches using MSR traces [1]. The traces are captured on 36 volumes of 179 disks located in 13 servers. They are composed of I/O requests, each specifying the timestamp, the server name, the disk number, the read/write type, the starting logical block address, the request size, and the response time. We choose six representative traces with different overwrite percentages to perform performance evaluation. Zero-think-time trace replays are conducted in the evaluation. Merging compactions are triggered during trace replaying.

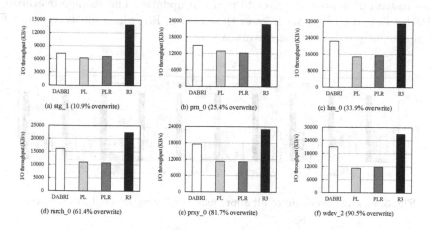

Fig. 4. I/O throughput of all approaches replaying selected MSR traces.

Figure 4 shows the I/O throughput of all approaches replaying selected MSR traces. R3 always has the highest I/O throughput for all selected traces. This is because no additional read, parity computation, and data compaction are involved in R3. DABRI significantly outperforms PL and PLR for all selected traces, particularly for the traces with high overwrite percentages. For example, Fig. 4(a) shows that DABRI outperforms the state-of-the-art PLR by 9.83% when the trace stg_1 with a overwrite percentage of 10.9% is replayed, whereas Fig. 4(f) shows that DABRI outperforms PLR by 77.41% when the trace $wdev\_2$ with a overwrite percentage of 90.5% is replayed. This behavior occurs because DABRI performs only a single write for the $r^{(th)}$ $(r \geq 2)$ partial writes on the same data; the higher the overwrite percentage of a trace, the larger is the advantage of DABRI.

PLR reserves a certain amount of disk space next to each parity chunk such that the parity updates can be performed in the reserved space to mitigate disk seeks. However, PLR and PL give similar I/O performance with less than 6.05% difference in I/O throughput. This is because both PLR and PL perform a time-consuming write-after-read for every partial write; the offsets of requests are randomly mapped to different chunks instead of a single chunk, and no significant seek overhead exists between PLR and PL.

5.2   Storage Overhead

We choose 10 representative traces in MSR traces to perform evaluation on storage overhead. Figure 5 shows the storage overhead of all approaches replaying selected MSR traces. R3 has the highest storage overhead replaying all selected traces, and its storage overhead is 3× and remains unchange regardless of traces. This is because R3 keeps three replicas for every data chunk, and uses in-place writes instead of log-based writes to perform updates. The storage overhead of DABRI, PL, and PLR is much lower than that of R3, which illustrates that EC can significantly reduce storage overhead.

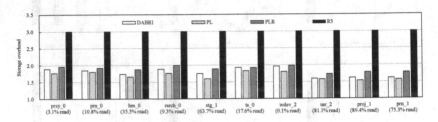

Fig. 5. Storage overhead of all approaches replaying selected MSR traces.

In Fig. 5, PL is the baseline, and has the lowest storage overhead under different traces. The storage overhead of PLR is higher than that of PL. This is because PLR reserves a certain amount of disk space next to each parity chunk. However, the storage overhead of PLR is not greater than 2×. This is because the parity deltas in the reserved space is merged to each parity chunk, and the unused reserved space is dynamically released. The storage overhead of DABRI is slightly higher than that of PL. This is because DABRI performs logging-based update for the first write on the same data. The storage overhead of DABRI is lower than PLR for most traces.

5.3   Degraded Read

When the data located on failure data nodes are read, erasure-coded system clients must recover the requested data by retrieving the available data and parity chunks. This procedure is called degraded read. We replay MSR traces to compare the degraded read performance for DABRI, PL, and PLR. We choose 10 representative traces in MSR traces to perform evaluation. We randomly delete a data chunk from the local file system on a data node, which is detected by the data node as a missing chunk, to trigger a degraded read.

Figure 6 shows the degraded read time of DABRI and PLR compared to PL, where PL is the base line. DABRI outperforms PL for most traces. This is because DABRI can read the latest data from parity nodes without decoding whenever possible. If the latest data are not stored on the parity nodes, then the client needs to read a subset of data and parity chunks on surviving nodes to recover requested data. PLR significantly outperforms PL for all selected traces, particularly for the write-dominant traces. This is because each parity chunk and its parity deltas can be sequentially retrieved.

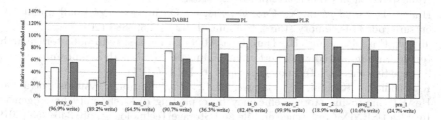

Fig. 6. Relative degraded read time of DABRI and PLR compared to PL.

6 Conclusion and Future Work

We have proposed DABRI, an efficient data delta based hybrid write scheme to support fast partial writes for erasure-coded storage systems. DABRI has appended the new data to the logs on data nodes and parity nodes for the first partial write. With hybrid writes, DABRI has successfully realized parallel I/O between data nodes and parity nodes, and avoided read degradation. Experimental results show that the DABRI has successfully improved the I/O throughput by up to 77.41% on real-world traces, compared with the sate-of-the-art.

Many modern storage solution make heavy usage of NVMe. NVMe offer much better latency and support unaligned access, the cost of an I/O operations is much lower. In the future, we provide a performance model taking into account communication, I/O latency for partial writes.

Acknowledgment. This work was supported in part by the China Postdoctoral Science Foundation under Grant No. 2021M690733, the Key-Area Research and Development Program of Guangdong Province under Grant 2019B010121001, and the National Natural Science Foundation of China under Grant No. 62072118.

References

1. Chan, J.C., Ding, Q., Lee, P.P., Chan, H.H.: Parity logging with reserved space: towards efficient updates and recovery in erasure-coded clustered storage. In: 12th USENIX Conference on File and Storage Technologies, pp. 163–176 (2014)
2. Chen, Y.L., Mu, S., Li, J., Huang, C., Li, J., Ogus, A., Phillips, D.: Giza: erasure coding objects across global data centers. In: 2017 USENIX Annual Technical Conference, pp. 539–551 (2017)
3. Ghemawat, S., Gobioff, H., Leung, S.T.: The google file system. In: Proceedings of the Nineteenth ACM Symposium on Operating Systems Principles, pp. 29–43 (2003)

4. Hu, Y., Cheng, L., Yao, Q., Lee, P.P., Wang, W., Chen, W.: Exploiting combined locality for wide-stripe erasure coding in distributed storage. In: 19th USENIX Conference on File and Storage Technologies, pp. 233–248 (2021)
5. Huang, C., et al.: Erasure coding in windows azure storage. In: 2012 USENIX Annual Technical Conference, pp. 15–26 (2012)
6. Huang, J., Xia, J., Qin, X., Cao, Q., Xie, C.: Optimization of small updates for erasure-coded in-memory stores. Comput. J. 62(6), 869–883 (2019)
7. Jin, C., Feng, D., Jiang, H., Tian, L.: RAID6L: a log-assisted raid6 storage architecture with improved write performance. In: 2011 IEEE 27th Symposium on Mass Storage Systems and Technologies, pp. 1–6. IEEE (2011)
8. Kadekodi, S., Rashmi, K., Ganger, G.R.: Cluster storage systems gotta have HeART: improving storage efficiency by exploiting disk-reliability heterogeneity. In: 17th USENIX Conference on File and Storage Technologies, pp. 345–358 (2019)
9. Li, X., Li, R., Lee, P.P., Hu, Y.: OpenEC: toward unified and configurable erasure coding management in distributed storage systems. In: 17th USENIX Conference on File and Storage Technologies, pp. 331–344 (2019)
10. Plank, J.S., Greenan, K.M., Miller, E.L.: Screaming fast Galois field arithmetic using Intel SIMD instructions. In: 11th USENIX Conference on File and Storage Technologies, pp. 299–306 (2013)
11. Shen, J., Zhang, K., Gu, J., Zhou, Y., Wang, X.: Efficient scheduling for multi-block updates in erasure coding based storage systems. IEEE Trans. Comput. 67(4), 573–581 (2017)
12. Shvachko, K., Kuang, H., Radia, S., Chansler, R.: The Hadoop distributed file system. In: 2010 IEEE 26th Symposium on Mass Storage Systems and Technologies, pp. 1–10. IEEE (2010)
13. Silberstein, M., Ganesh, L., Wang, Y., Alvisi, L., Dahlin, M.: Lazy means smart: reducing repair bandwidth costs in erasure-coded distributed storage. In: Proceedings of International Conference on Systems and Storage, pp. 1–7 (2014)
14. Subedi, P., Huang, P., Young, B., He, X.: FINGER: a novel erasure coding scheme using fine granularity blocks to improve Hadoop write and update performance. In: 2015 IEEE International Conference on Networking, Architecture and Storage, pp. 255–264. IEEE (2015)
15. Wang, Y., Pei, X., Ma, X., Xu, F.: TA-Update: an adaptive update scheme with tree-structured transmission in erasure-coded storage systems. IEEE Trans. Parallel Distrib. Syst. 29(8), 1893–1906 (2017)
16. Wei, B., et al.: A self-tuning client-side metadata prefetching scheme for wide area network file systems. Sci. China Inf. Sci. 65(3), 1–17 (2021). https://doi.org/10.1007/s11432-019-2833-1
17. Wei, B., Xiao, L., Zhou, B., Qin, G., Yan, B., Huo, Z.: Fine-grained management of I/O optimizations based on workload characteristics. Front. Comput. Sci. 15(3), 1–14 (2021)
18. Ye, L., Feng, D., Hu, Y., Wei, X.: Hybrid codes: flexible erasure codes with optimized recovery performance. ACM Trans. Storage 16(4), 1–26 (2020)
19. Zhou, T., Tian, C.: Fast erasure coding for data storage: a comprehensive study of the acceleration techniques. ACM Trans. Storage 16(1), 1–24 (2020)

BDCuckoo: an Efficient Cuckoo Hash for Block Device

Xianqi Zheng, Jia Ma, Yubo Liu, and Zhiguang Chen[✉]

Sun Yat-sen University, Guangzhou 510275, People's Republic of China
{zhengxq27,majia5}@mail2.sysu.edu.cn,
{yubo.liu,zhiguang.chen}@nscc-gz.cn

Abstract. Hash is widely used in various storage systems due to its excellent insertion and search performance. However, existing hash designs are not friendly for block devices because they will generate a lot of random small I/Os, which will significantly reduce the I/O efficiency on block devices. This paper proposes BDCuckoo (Block Device Cuckoo) hash, a I/O-optimized Cuckoo hash for block device. BDCuckoo reduces the amount of I/Os during the slot detection by limiting the location where the element may be stored on the hash table. Unlike the traditional cuckoo hash that triggers an disk I/O for each detection, BDCuckoo hash loads the possible slots of the target element into DRAM in a single large disk I/O. The paper also shows a use case that uses BDCuckoo to optimize a large directory index in the EXT4 file system. The evaluation shows that BDCuckoo hash outperforms the traditional Cuckoo hash for all YCSB workloads been tested and has a 2.64 times performance improvement at most for workload D with the load factor of 0.7. In the use case, the directory index of BDCuckoo-optimized file system achieves 1.79 times performance improvement for *stat* command and 1.64 times performance improvement for *rm* command.

Keywords: Hash · I/O · Block device · File system · Large directory

1 Introduction

Hash is a widely used index structure in storage systems. Compared with other indexes (e.g., tree, list), one of the important advantages of hash is the constant time complexity in element insertion and searching. A common usage is to use hash to index the data cached in DRAM. However, existing hashes are not friendly for block devices because they will trigger a lot of random small I/Os, which makes block devices unable to benefit from hash. Cuckoo hash is a mature and efficient hash used in a variety of production environments. The core design of Cuckoo hash is to map the element to multiple slots using multiple hash algorithms. When a hash conflict occurs, the element in the current slot is evicted to its other hash position. By this design, Cuckoo hash can ensure high search efficiency, because each search only needs to detect a constant number of slots

© IFIP International Federation for Information Processing 2022
Published by Springer Nature Switzerland AG 2022
C. Cérin et al. (Eds.): NPC 2021, LNCS 13152, pp. 183–194, 2022.
https://doi.org/10.1007/978-3-030-93571-9_15

(equal to the number of hash algorithms). However, the element eviction in insertion and detection in searching will generate multiple random small I/Os, which significantly reduce the efficiency of Cuckoo hash on block devices.

The goal of this work is to improve the efficiency of Cuckoo hash on block devices. According to our observations, if two small I/Os are within a certain distance of each other on a block device, then it is more efficient to aggregate them into one continuous large I/O than to trigger two small I/Os. Base on this motivation, this paper designs BDCuckoo (Block Device Cuckoo). BDCuckoo limits multiple possible locations to a given distance. According to our practical experience, the distant is set to 20 blocks. This design allows BDCuckoo can load all possible locations of the target key in a single, fixed-size disk I/O, thereby improving the disk I/O efficiency of location detection in Cuckoo hash on block devices.

To further demonstrate the advantage of BDCuckoo, we leverage it to optimize the large directory index in the traditional file system (EXT4). The original EXT4 [1] uses a HTree [2] to index the entries in a directory. HTree is a B-tree with limited height whose logarithmic time complexity limits the search performance in the case of large directory. However, the poor performance of traditional hashes on block devices, they are not suitable to be used to solve the large directory problem in file systems. We replaced the HTree by BDCuckoo to achieve constant time directory lookup.

The contributions of this work are as follows:

1) It designs and implements BDCuckoo. BDCuckoo improves the I/O efficiency of location detection by limiting the distant of multiple possible locations of the element.
2) It optimizes the performance of EXT4 in the case of large directory by using BDCuckoo to index the items in a directory, thereby increasing the efficiency of directory lookup to a constant time complexity.
3) It evaluates BDCuckoo with YCSB and the BDCuckoo-optimized EXT4 with a variety of common file system operations. The results show that BDCuckoo outperforms Cuckoo for all YCSB workloads been tested on block devices and has a 2.64 times performance improvement at most; the performance of BDCuckoo-optimized EXT4 has advantages over traditional EXT4 in some common file system operations, for example, its performance is more than 1.79 times that of the traditional EXT4 in stat operation in the case of large directory.

The rest of the paper is organized as follows. Section 2 describes the background and motivations. Section 3 presents the BDCuckoo design and detailed algorithms. Section 4 discusses a use case for BDCuckoo. Section 5 shows the evaluation results. Section 6 introduces the related works and Sect. 7 gives a conclusion.

2   Background and Motivation

2.1   Hash on Block Devices

Block devices (e.g., HDD and SSD) are widely used for large-scale data storage, they store data in fixed-size blocks. Although most block devices have a high sequential throughput, the random I/O performance of block devices is low [3]. For example, the magnetic hard disk usually becomes the performance bottle-neck for disk-based applications because of its limited number of IOPS. Even enterprise class hard disks can hardly deliver more than 300 IOPS [4]. As for the solid state drive, although it promises an order of magnitude higher IOPS than the magnetic hard disk, its random access performance is still limited compared with DRAM [5,6].

According to the characteristics of block devices, system designers usually use tree structure to index data on block devices, so as to avoid the shortcomings of block devices on random small I/O as much as possible. Compared with tree structures, hash provides constant time complexity for the insert, lookup, and delete operations on average. Hash is usually used in scenarios that are sensitive to access latency. However, there is difficult to get benefit from hash on block devices because the elements will be randomly distributed in the table. In addition, in order to resolve hash conflicts, hash algorithms usually need to perform multiple detections during insertion/search. These behaviors of hashes will bring a lot of random small I/O, making them unsuitable for block devices.

2.2   Cuckoo Hash

Cuckoo hash [7,8] is a hash collision resolution algorithm that is widely used hash in many systems. Basically, cuckoo hash uses multiple hash functions and each element in cuckoo hash is mapped to multiple possible buckets in the hash table. Although an element can be stored in at most one of its possible buckets at a time, it can move between all its possible buckets. Moreover, the hash buckets in cuckoo hash are usually multi-way set associative, i.e., each bucket has B "slots". $B = 4$ is a common value in practice. A lookup for an element proceeds by computing all possible buckets of the element and check all of the slots within each of those buckets. When inserts an element X into the table, if one of the possible buckets has empty slot, then the element is inserted into that bucket and the insertion operation finishes. If none of the possible buckets has empty slot, the algorithm randomly chooses a slot from the candidate buckets and kicks out the occupant Y and inserts X into that slot, then relocates Y to its own alternate location, possibly displacing another element, and so on, until a maximum number of displacements is reached, which indicates a insertion failure.

2.3   Motivation

The possible location detection in Cuckoo hash is an important factor that limits its performance on block devices. For example, a single insertion operation may result in many evictions and re-insertions operations followed, and a lookup operation in cuckoo hash requires checking all the possible locations of the target element. Considering that the sequential I/O performance of block device is much better than its random I/O performance, the motivation of this work is to restrict all possible hashing locations of a specific element to be distributed in a physically contiguous space and use a sequential big I/O that covers all possible hashing locations of the element to replace the multiple discrete small I/O in the lookup and insertion process. We believe such strategy can reduce the disk I/O overhead effectively.

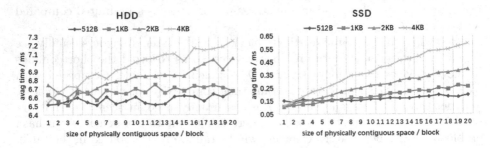

Fig. 1. The disk IO experiment results.

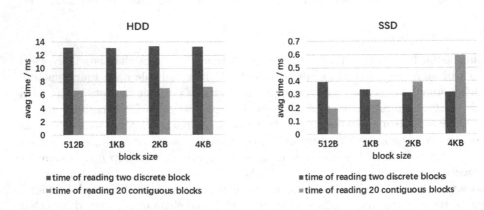

Fig. 2. The time cost comparison.

We use two experiments to verify our motivation. Let's assume each element in cuckoo hash table T has two possible hashing locations. We first initialize the size of the physically contiguous space to a single block, then increase it by a block each time and measure the sequential reading time. We do our experiments within four block sizes: 512 B, 1 KB, 2 KB and 4 KB. Figure 1 shows the sequential reading time from 1 block to 20 blocks. We can see that as the size of the physically contiguous space increases, the sequential reading time increases on both HDD and SSD. However, it is noteworthy that sequential reading time on HDD increases very slowly within all four block sizes, while that of SSD increases rapidly with the block size of 2 KB and 4 KB.

We compare the time of reading 20 blocks sequentially and the time of reading two discrete blocks randomly within all four block sizes. The results are showed in Fig. 2. As shown in Fig. 2, we can see that the sequential access of 20 blocks outperforms the random access of two discrete blocks so much better within all four block sizes on HDD. The results on SSD are some kind of different. The sequential I/O performs better with the block sizes of 512 B and 1 KB, but the random I/O performs better in the 2 KB and 4 KB block size situations.

In summary, these experiments show that the strategy of using a sequential big I/O to replace multiple discrete small I/O can reduce the disk I/O overhead on HDD, and also works on SSD if we use a small block size such as 512 B and 1 KB.

3   BDCuckoo Design

The structure of BDCuckoo is the same as the traditional Cuckoo hash. Each hashing location of BDCuckoo is 512 bytes and corresponds to a block. There are two hash functions (CityHash [9] and MurmurHash [10]) in the prototype of BDCuckoo, which means that each element has two possible locations. BDCuckoo restricts the two possible hashing locations of element X to be distributed in physically adjacent 20 blocks. The two hashing locations in BDCuckoo are determined as follows: first, use CityHash to get the first hash value, namely $H_1(X)$, which indicates the index of the first hashing location in the hash table. Next, use MurmurHash to get the second hash value $H_2(X)$ and divide $H_2(X)$ by 20, then get the modulo as bias b and take the sum of $H_1(X)$ and bias b as the index of the second hashing location in the hash table. In this way, BDCuckoo ensures that the two possible hashing locations of a specific element is distributed in physically adjacent 20 blocks.

3.1   Main Operations in BDCuckoo

Insertion. The insertion process of BDCuckoo is different from that of the traditional cuckoo hash. When inserts an element X, BDCuckoo firstly uses CityHash to calculate the index of the first possible hashing location of element X, namely $H_1(x)$. Then it reads the physically adjacent 20 blocks from disk starting with index $H_1(x)$. The disk I/O unit in BDCuckoo is 20 blocks, and this allows BDCuckoo to load all possible positions of the element to the memory with one large I/O. The first possible hashing location of element X is the first block of the 20 in-memory blocks, and the index of the second possible hashing location in the 20 in-memory blocks is the above-mentioned bias b, which can be got by the MurmurHash and a modular operation. Then BDCuckoo checks the two possible hashing locations in memory. If there is an empty location, then the element X is placed into the location and then the location is written back to disk, and the insertion operation completes. If both the two possible hashing

---

Algorithm 1: function INSERT(x)

```
 1  for i ← 0 to MAX_ATTEMPTS by 1 do
 2  │   index1 = CityHash(x);
 3  │   Read(buf, index1*blockSize, 20*blockSize);
 4  │   index2 = MurmurHash(x) mod 20;
 5  │   if buf[0] is empty then
 6  │   │   Swap(x, buf[0]);
 7  │   │   write buf[0] back to disk;
 8  │   │   return true;
 9  │   else
10  │   │   if buf[index2] is empty then
11  │   │   │   Swap(x, buf[index2]);
12  │   │   │   write buf[index2] back to disk;
13  │   │   │   return true;
14  │   │   else
15  │   │   │   index = RandPick(0, index2);
16  │   │   │   Swap(x, buf[index]);
17  │   │   │   write buf[index] back to disk;
18  │   │   end
19  │   end
20  end
21  return false
```

---

locations are occupied, then the algorithm chooses one of the locations randomly and kicks out the current occupant Y. Next places X into the location and then writes the location back to disk. The same procedure is executed for element Y. The insertion and the eviction operations continue until no collision happens or a maximum number of displacements is reached. The maximum number of displacements is set to 50, which is empirically derived.

Algorithm 1 describes the process of the insertion operation. In Algorithm 1, the *MAX_ATTEMPTS* denotes the maximum number of displacements in a single insertion operation. The *Read* function reads the physically contiguous 20 blocks from disk into the in-memory buffer *buf*. The *Swap* function swaps the values of the given variable x and variable y. The *RandPick* function randomly chooses one variable from a set of given variables.

Lookup. The lookup process of BDCuckoo is simple. When looks for an element X, the lookup process firstly uses the CityHash to calculate the first possible hashing location, namely $H_1(x)$. Then it reads the physically adjacent 20 blocks from disk starting with index $H_1(x)$. Next, BDCuckoo checks the two possible hashing locations of element X within the 20 in-memory blocks and returns the target data if found. Otherwise, returns *NULL*.

Update and Delete. The update operation and delete operation are similar to the lookup operation. BDCuckoo firstly reads the physically adjacent 20 blocks into memory, then checks the two possible hashing locations of element X within the 20 in-memory blocks, updates or deletes the target data if found.

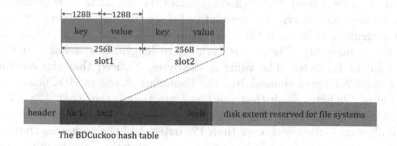

Fig. 3. The on-disk layout.

Rehash. As the number of files within a directory grows continuously, the hash table may become so full that either a single insertion operation may incur many evictions and re-insertions followed or an insertion failure happens. Therefore, we set a threshold to the load factor of the hash table. As long as the load factor of the hash table reaches the threshold, the rehashing operation will be triggered. Considering that each element in BDCuckoo has two possible hashing locations, we set the rehashing threshold to 0.5. And our evaluations show that the possibility of insertion failure is low, almost negligible when the threshold is set to 0.5.

4   Case Study: Optimization of Large Directory Lookup

In this section, we introduce a use case for BDCuckoo. Directory Lookup is a frequent operation in file system. However, finding the target entry in a large directory is expensive, because current file systems mostly use complicated tree structures to index the metadata of all files and directories and the efficiency of the tree index will decrease as the amount of data grows. Take the famous EXT4 [1] file system as a example. The EXT4 file system uses a HTree [2] structure to index the metadata. HTree is a two-level tree structure, with the root node in the first level and the leaf nodes in the second level. Every leaf node contains many entries, and each entry corresponds to a file or a directory. Entries in the same leaf node are ordered. The challenge of large catalog indexing exists in two parts. First, the total amount of metadata can be too big to be fully cached. Second, the tree-based indexing structures may become so deep and the on-disk path lookup needs to access each level of the hierarchical structure by sequence, which is very costly. To deal with the above-mentioned problem, we use BDCuckoo to manage the metadata. BDCuckoo is a flattened structure and it can avoid the expensive hierarchical path lookup overhead.

We divide the disk into two parts, one is for BDCuckoo and the other is for the local file systems. There is a 512 bytes header in BDCuckoo, which records information about BDCuckoo such as the total size of the hash table and the hash functions been used. BDCuckoo organizes the metadata as key-value pairs. Each key-value pair stores the metadata of a specific file or directory, the key field is a filename string and the value field contains the Inode [11] attributes of the file or directory. The key is variable in length, but it has a maximum length limit of 128 bytes. The value is 128 bytes. Noticed that the filename in Linux can be 255 bytes at most but the filenames stored in BDCuckoo have a limitation of 128 bytes, such that the filenames that longer than 128 bytes can not be stored in BDCuckoo. However, it is noteworthy that most of the filenames in the daily workloads are shorter than 128 bytes, and the filenames that exceed 128 bytes can be stored into the local file systems directly. Each hashing location of BDCuckoo is 512 bytes and contains two slots, each of which is 256 bytes and contains a metadata key-value pair. The key-value pairs in the two slots within the same hashing location have the same hash value. The on-disk layout is showed in Fig. 3.

5   Evaluation

This section evaluates the performance of BDCuckoo and the BDCuckoo-optimized EXT4. All the results are collected from a Linux server equipped as follows:

Linux   CentOS 7.6, 64-bit version
CPU   Intel® Xeon® Gold 6230N CPU
DRAM   DDR4 SDRAM, 187 GB
HDD   TOSHIBA AL14SEB060NY, 600 GB
SSD   DELL SCV1DL57, SATA 3.2, 6.0 Gb/s, 480 GB

The evaluations were run on the cold cache when we do not specify. We cleaned the memory after each experiment to make the cache clean.

5.1   Performance of BDCuckoo

We use YCSB [12] benchmark to evaluate the performance of BDCuckoo and compare it with the normal cuckoo hash. We tested four different YCSB workloads: workload A(50% read + 50%update), workload B(95%read + 5%update), workload C(100%read) and workload D(95%read + 5%insert). Moreover, We repeat our experiments within three different load factors: 0.3, 0.5, 0.7. The experiment results are showed in Fig. 4. As shown in Fig. 4, BDCuckoo outperforms the normal cuckoo in all situations and has a 2.64 times performance improvement at most for workload D with the load factor of 0.7.

5.2   Performance of BDCuckoo-Optimized EXT4

As mentioned in Sect. 4, BDCuckoo is an efficient directory indexing structure and can be used to manage the metadata. Here we implement a fuse-based [13] prototype named BDCuckoo-fuse and compare it with the EXT4 file system. Five frequent used metadata operations are selected for our evaluation: creating files, get attributes of files, change the mode of files, change the owner of files and removing files. These five metadata operations correspond to five Linux terminal commands: *touch, stat, chmod, chown, rm.*

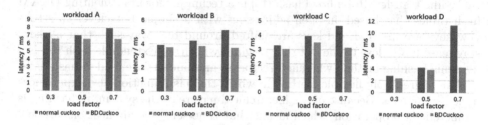

Fig. 4. The YCSB results.

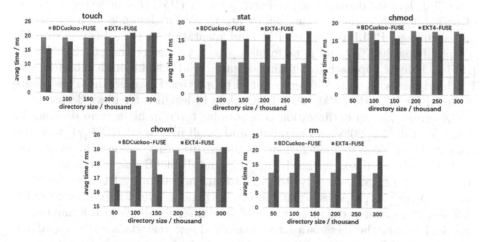

Fig. 5. Performance of metadata operations.

We tested the five Linux terminal commands under six different directory sizes (number of entry), ranged form 50 thousands to 300 thousands. All five commands were executed 20000 times under each directory size. For each command, We recorded the duration of each execution and took the average time as the time cost of executing the command. The results are showed in Fig. 5.

As shown in Fig. 5, BDCuckoo implements a substantial speedup for the *stat* command and *rm* command under all six different directory sizes. It achieves 1.79

times performance improvement for *stat* command and 1.64 times performance improvement for *rm* command on average. For the other three commands, the time cost of BDCuckoo-fuse and EXT4-fuse is similar.

6   Related Work

Some research has focused on optimizing hash for different kinds of hardware medias. For example, CLHT [14] places each hash-table bucket on a single cache line and performs in-place updates so that operations complete with at most one cache-line transfer. Robinhood hash [15] is a technique for implementing DRAM hash tables that based on open addressing with a simple but clever twist: As new keys are inserted, old keys are shifted around in a way such that all keys stay reasonably close to the slot they originally hash to. Extendible hash [16] is a dynamic hashing method wherein directories, and buckets are used to hash data. It is an aggressively flexible method in which the hash function also experiences dynamic changes. Because of the hierarchical nature of the system, re-hashing is an incremental operation, which means that time-sensitive applications are less affected by table growth than by standard full-table rehashes. Meanwhile, there are also some other works been carried out to deploy hash-based structures on other hardware medias. For example, CCEH [17] is a cacheline-conscious extendible hashing deployed on persistent memory (PM) that makes effective use of cachelines while guaranteeing failure-atomicity for dynamic hash expansion and shrinkage. PCLHT [18] is a PM implementation of the CLHT hash table by modifying 30 LOC and it performs up to 2.4× better than CCEH. DASH [19] is a holistic approach to building dynamic and scalable hash tables on real PM hardware with all the aforementioned properties. However, these hashes are designed for DRAM or PM, while BDCuckoo is designed for block devices.

Directory lookup optimization is a popular topic in file system design. For example, TableFS [20] stores metadata and small files in LevelDB [21] to accelerate multiple file system operations, including directory lookup. However, LevelDB is designed based on LSM Tree [22] and TableFS does not take benefit from hash. DLFS [23] directly stores metadata and data by hash on the block device. Otter [24] proposes a metadata management layer for existing local file systems. It separates the metadata based on their locations in the namespace tree and hashes the metadata into some fixed-size metadata buckets (MDB). Although DLFS and Otter speed up directory lookup by using hash to index metadata/data, they do not realize the potential of hash on block devices.

7   Conclusion

This paper proposes an efficient cuckoo hash for block device named BDCuckoo. BDCuckoo uses a sequential big I/O to replace multiple discrete small I/O in normal cuckoo hash. It evaluates the performance of BDCuckoo by YCSB benchmark and compare BDCuckoo with the traditional Cuckoo hash. The results show that BDCuckoo has a significant performance improvement for the normal cuckoo hash. Further more, it optimizes EXT4 in the case of large directory lookup by using BDCuckoo to replace the HTree as the entry index and the evaluation shows that BDCuckoo-optimized EXT4 outperforms original EXT4 in many common file system operation in the case of large directory.

Acknowledgments. We thank the reviewers for their insightful feedback to improve this paper. This work was supported by National Key R&D Program of China (2018YFC1406205), National Natural Science Foundation of China (No. 61872392,61832020), Zhejiang Lab (NO. 2021KC0AB04), Key-Area Research and Development Program of Guangdong Province (2019B010107001), Guangdong Natural Science Foundation (2018B030312002), Pearl River S & T Nova Program of Guangdong (201906010008) and the Major Program of Guangdong Basic and Applied Research (2019B030302002).

References

1. Mathur, A., Cao, M., Bhattacharya, S., Dilger, A., Tomas, A., Vivier, L.: The new ext4 filesystem: current status and future plans. In: Proceedings of the Linux symposium, vol. 2, pp. 21–33. Citeseer (2007)
2. Phillips, D.: A directory index for EXT2. In: Annual Linux Showcase & Conference (2001)
3. Chung, L., Gray, J., Horst, R., Worthington, B.: Windows 2000 disk IO performance (2000)
4. Meister, D., Brinkmann, A.: dedupv1: improving deduplication throughput using solid state drives (SSD). In: 2010 IEEE 26th Symposium on Mass Storage Systems and Technologies (MSST), pp. 1–6. IEEE (2010)
5. Narayanan, D., Thereska, E., Donnelly, A., Elnikety, S., Rowstron, A.: Migrating server storage to SSDs: analysis of tradeoffs. In: 2009 Proceedings of the 4th ACM European Conference on Computer Systems, pp. 145–158 (2009)
6. Myers, D.D.S.: On the use of NAND flash memory in high-performance relational databases. Ph.D. dissertation, Massachusetts Institute of Technology (2008)
7. Devroye, L., Morin, P.: Cuckoo hashing: further analysis. Inf. Process. Lett. **86**(4), 215–219 (2003)
8. Li, X., Andersen, D.G., Kaminsky, M., Freedman, M.J.: Algorithmic improvements for fast concurrent cuckoo hashing. In: 2014 Proceedings of the 9th European Conference on Computer Systems, pp. 1–14 (2014)
9. Pike, G.: CityHash: fast hash functions for strings. Stanford University class slides, October 2012 (2012)
10. Yamaguchi, F., Nishi, H.: Hardware-based hash functions for network applications. In: 2013 19th IEEE International Conference on Networks (ICON), pp. 1–6. IEEE (2013)
11. Aneesh Kumar, K.V., Cao, M., Santos, J.R., Dilger, A.: Ext4 block and inode allocator improvements. In: Linux Symposium, vol. 1 (2008)
12. Barata, M., Bernardino, J., Furtado, P.: YCSB and TPC-H: big data and decision support benchmarks. In: 2014 IEEE International Congress on Big Data, pp. 800–801. IEEE (2014)
13. Vangoor, B.K.R., Tarasov, V., Zadok, E.: To fuse or not to fuse: performance of user-space file systems. In: 2017 Proceedings of USENIX Conference on File and Storage Technologies (FAST), pp. 59–72 (2017)

14. David, T., Guerraoui, R., Trigonakis, V.: Asynchronized concurrency: the secret to scaling concurrent search data structures. ACM SIGARCH Comput. Architect. News **43**(1), 631–644 (2015)
15. Marcus, R., et al.: Benchmarking learned indexes. arXiv preprint arXiv:2006.12804 (2020)
16. Fagin, R., Nievergelt, J., Pippenger, N., Strong, H.R.: Extendible hashing-a fast access method for dynamic files. ACM Trans. Database Syst. **4**(3), 315–344 (1979)
17. Nam, M., Cha, H., Choi, Y.-R., Noh, S.H., Nam, B.: Write-optimized dynamic hashing for persistent memory. In: 2019 Proceedings of USENIX Conference on File and Storage Technologies (FAST), pp. 31–44 (2019)
18. Lee, S.K., Mohan, J., Kashyap, S., Kim, T., Chidambaram, V.: RECIPE: converting concurrent dram indexes to persistent-memory indexes. In: 2019 Proceedings of the 27th ACM Symposium on Operating Systems Principles, pp. 462–477 (2019)
19. Lu, B., Hao, X., Wang, T., Lo, E.: Dash: scalable hashing on persistent memory. Proc. VLDB Endow. **13**(8), 1147–1161 (2020)
20. Ren, K., Gibson, G.: TABLEFS: enhancing metadata efficiency in the local file system. In: 2013 Proceedings of USENIX Technical Conference (ATC), pp. 145–156 (2013)
21. Dent, A.: Getting Started with LevelDB. Packt Publishing Ltd. (2013)
22. O'Neil, P., Cheng, E., Gawlick, D., O'Neil, E.: The log-structured merge-tree (LSM-tree). Acta Informatica **33**(4), 351–385 (1996)
23. Lensing, P.H., Cortes, T., Brinkmann, A.: Direct lookup and hash-based metadata placement for local file systems. In: Proceedings of the 6th International Systems and Storage Conference, pp. 1–11 (2013)
24. Liu, Y., Li, H., Lu, Y., Chen, Z., Zhao, M.: An efficient and flexible metadata management layer for local file systems. In: 2019 IEEE 37th International Conference on Computer Design (ICCD). IEEE, pp. 208–216 (2019)

A Two Tier Hybrid Metadata Management Mechanism for NVM Storage System

Tao Cai(✉) , Pengfei Gao(✉), Fuli Chen, Dejiao Niu , Fei Wang, Yueming Ma, and Lei Li

Jiangsu University, Zhenjiang 212013, China

Abstract. NVM is an effective method for improving the performance of storage systems. Meanwhile, metadata management efficiency has a significant impact on the I/O performance of storage systems. But the current metadata management strategy is not efficient and will decrease the endurance of NVM devices. We design a two tier hybrid metadata management mechanism. Based upon the intrinsic characteristics of NVM devices and metadata management, the metadata management is decomposed instead of concentrating in the file system and redistributed between the file system and the device driver. A hybrid distributed directory tree, a dual-zone cache management algorithm and a multi-skiplists index embedded in driver are designed to short the I/O system software stack for accessing and managing metadata, and improve the concurrency of metadata management especially in computer with multi-core processor. Then, the prototype is implemented on a computer system equipped with Intel Optane DC based on the PMEM to evaluate with several common application workloads. The results demonstrate that it can improve the IOPS by 76.9% and 26.9% compared with EXT4 and NOVA on NVM respectively, and it has better scalability for the change of the number of files and threads.

Keywords: NVM · Storage system · Metadata management

1 Introduction

The emergence of non-volatile memory (NVM), especially the commercialization of Intel Optane DC [1], has provided a good foundation for solving the storage wall problem of computer systems. However, the high-speed read and write capabilities of NVM have brought great challenges to the I/O system software stack. Then it becomes an important bottleneck that affects the performance of storage and access data. Related research shows that the overhead of I/O system software accounts for the 94% total overhead for NVM storage systems [2]. The file system is the important I/O system software. The metadata is used to organize and manage the basic information about all files and directories in file systems. However, the amount of metadata is huge and they cannot be stored in the main memory. When using the existing I/O system software stack management, there are some limitations such as repeated metadata transmission between the storage device and main memory, multi-layer cache, etc. It is difficult to effectively take advantage of the fast read and write speed of the NVM device.

© IFIP International Federation for Information Processing 2022
Published by Springer Nature Switzerland AG 2022
C. Cérin et al. (Eds.): NPC 2021, LNCS 13152, pp. 195–213, 2022.
https://doi.org/10.1007/978-3-030-93571-9_16

Currently, the metadata [3] is generally stored and managed centrally by the file system using the index tree or the hash-based method. The index tree can well reflect the hierarchical structure of files and directories in file system, but all directories in the access path need to be accessed one by one, which leads to problems such as high time overhead. Hash-based methods can quickly find target file, but it is difficult to efficiently perform list operations. At the same time, when the number of files is huge, the calculation overhead of the hash function and the rate of hash conflict will increase sharply. This affects the efficiency of metadata management. In addition, the lock is needed in index trees and hash-based methods to ensure the consistency of metadata. In a multi-core computer, a large number of lock operations will bring huge additional time overhead. Therefore, how to change the centralized management of metadata by the file system and to study an efficient metadata management mechanism for the characteristics of NVM devices is an important issue.

This article first analyzes the existing metadata management mechanism. On this basis, aiming at the problems of large memory usage, high CPU resource consumption, and long I/O system software stack existing in the current metadata management mechanism, this paper presents the structure of a two tier hybrid metadata management mechanism, a hybrid distributed directory tree, a dual-zone cache management algorithm, and a multi-skiplists index embedded in driver. A prototype is implemented on a computer system equipped with Intel Optane DC to evaluate with several common application workloads.

The main contributions of our work can be summarized as follows:

1) In response to the challenges brought by NVM devices to the I/O system software stack, the metadata management is decomposed and redistributed between the file system and the device driver. Thereby it can short the I/O system software stack for accessing and managing metadata, and improving the efficiency of metadata management.

2) In order to improve the efficiency of metadata management, a hybrid distributed directory tree is designed to decompose the directory tree of file system. The traditional directory tree is splitted into a centralized metadata layer and a dispersed metadata layer. The centralized metadata layer is stored in DRAM and only contains the first N levels of directory trees, which can reduce the time overhead required to calculate the hash value and improve the efficiency of metadata management. The dispersed metadata layer is stored in the NVM and contains the remaining layers of the directory tree, the optimized skiplist is used to manage.

3) A dual-zone cache management algorithm is designed. A fixed-size window is used to manage the cache by the frequency of metadata access in the dispersed metadata layer, and the high-frequency metadata should be transferred to the cache located in DRAM, which can reduce the additional time and space overhead of cache management and improve the speed of accessing metadata in the dispersed metadata layer.

4) A multi-skiplists index embedded in driver is designed. In the NVM device driver, several skiplists are combined to manage the dispersed metadata layer stored in the NVM, which can improve the concurrency efficiency of metadata management, reduce the rate of access conflicts, and short the I/O software stack for reading

and writing metadata. The concurrent access performance of metadata can be improved for computer with multi-core processor.

5) Based on the storage device of Intel Optane DC Persistent Memory, the prototype of a two-tier hybrid metadata management mechanism for NVM named HSNVMMS is implemented. Filebench is used to simulate the I/O characteristics of typical applications and evaluate the throughput with the number of files and threads changed. EXT4 and NOVA are used to compare, the test results show that, HSNVMMS can increase the maximum IOPS value by 76.9% compared with EXT4, and by 26.9% compared with NOVA.

2 Related Works

There are many researches on NVM such as NVM file system, NVM storage system and metadata management mechanism.

2.1 NVM File System

BPFS is a new file system for NVM devices [4]. BPFS divides file into two types based on whether the size of file exceeds 8 bytes. If the data file is less than 8 bytes, hardware primitive operation instructions is used directly perform in-situ update. Otherwise, CoW (Copy on Write, CoW) should be used. However, only the data that has not been updated is copied instead of the complete copy, because there is no need to copy the data that is about to be rewritten. BPFS provides atomic write operations for fine-grained data, and uses short-term shadow paging to ensure the consistency of metadata and data. The difference compared with the traditional shadow paging file system is that the update of the traditional shadow paging file system triggers a cascaded copy-on-write operation from the modified location to the root location of the file system tree. When the root location of the file system is updated, the change is only submitted. MMU (Memory Management Unit, MMU) and TLB table is used to realize the conversion of virtual address to physical address in SCMFS [5], which reduces the time overhead of the file system. And SCMFS shares the memory bus bandwidth, CPU cache and TLB (Translation Lookaside Buffer, TLB) table, and maps it to the virtual process space, which greatly simplifies the file reading and writing process. PMFS is the lightweight POSIX file system for NVM [6], by removing the Block Layer, and using the byte addressability of PM to use atomic in-place updates (Atomic In-place Updates), fine-grained logging (Logging at Cache-line Granularity) and the combination of the Copy On Write strategy optimizes data consistency. At the same time, PMFS also optimizes Mmap I/O. In traditional file system implementation, the use of Memory-mapped I/O requires that the pages in the storage device are first copied to DRAM. PMFS optimizes Mmap I/O, so that PM memory pages are directly mapped to the address space of the user mode application. NOVA is the log structure of mixed volatile and non-volatile memory, POSIX file system for NVM [7, 8]. It uses a separate storage method for log and index, stores Log in a slower NVM device, and stores the index in fast DRAM. At the same time, the Radix tree is used to implement the index and improve the efficiency of the query. NOVA stores log and file data to NVM, and stores the index by constructing a Radix tree structure in DRAM to speed up the search operation. Each index node in NOVA has its own log, which allows concurrent updates across files without synchronization operations. This structure enables high concurrency of file access. NVMCFS is a new file system for hybrid storage system that combines NVM devices and SSDs [9]. The head-tail layout and space management based on two layer radix-tree is provided to unify logic space between two type NVM devices. The dynamic file data distributed strategy and buffer for an individual file are used to speed up the access response and improve I/O performance.

2.2   NVM Storage System

DDBL is a hybrid memory prototype of NVM and DRAM [10]. Since NVM devices can not only form a mixed memory with DRAM, but also form a mixed external memory with HDD and SSD, DRAM can be used for caching NVM to reduce the wear and tear of NVM, and proposed an efficient page-based hybrid memory storage scheme is proposed by using a double dynamic bucket linked list to allocate and manage NVM space and age-based lazy caching strategy to manage DRAM buffer. Hibachi is a cooperative hybrid cache strategy for NVM and DRAM [11]. In order to improve the cache hit rate, Hibachi separates the read cache from the write cache by designing different management mechanisms. A dynamic page adjustment mechanism is used to flexibly adjust the sizes of clean cache and dirty cache and adapted to different operating loads. LosPem is a new log structure framework based on persistent memory composed of NVM devices and DRAM [12]. Firstly, it uses an efficient hash index linked list to maintain log content and reduces the overhead of log content retrieval. Secondly, LosPem decouples transactions into two asynchronous steps and creates a write buffer on the DRAM to handle frequent data writes to improve transaction throughput. Strata is a cross-media file system for several that can take leverages the strengths of one storage media to compensate for weaknesses of another [13]. Strata can provide performance, capacity, and a simple synchronous I/O model all at once, while having a simpler design than that of file systems constrained by a single storage device. The core of Strata lies in a log-structure approach with a novel split of responsibilities among user mode, kernel and storage layer that separates the concerns of scalability and high-performance persistence from storage layer management. Ziggurat is a tiered file system that combines NVM, DRAM and disk [14]. Ziggurat steers incoming writes to NVM, DRAM, or disk depending on application access patterns, write size, and the likelihood that the application will stall until the write completes. It profiles the application's access stream online to predict the behavior of individual writing. Meanwhile, Ziggurat migrates the cold file data from NVM to disks. Ziggurat coalesces data blocks into large, sequential writes to fully utilize disk bandwidth.

2.3   Metadata Management Mechanism

HIFFS is a new file system designed for NAND [15]. Different with the index tree for metadata in traditional file systems, Hash is used to speed up directory lookup performance. But the space utilization is low, because even if the hash item is empty, a fixed space must be allocated for the hash table. The hierarchical indexing strategy is used to index file data. Taking into account the remote update in the flash memory, if a data page is updated, its direct index file should also be updated. The update of the direct index results in the update of a single indirect index, which will result in a chain of recursive updates. Obviously, this is unacceptable for flash memory. They proposed a hybrid indexing scheme based on hash index, which uses hash to manage directory index to improve search efficiency. At the same time, according to the different file size, the adaptive method is used to select different file space indexing strategies. When using Varmail workload, the throughput of HIFFS is increased by 20 times, 22 times and 35 times than Ext2, Ext4 and Btrfs respectively. The new directory tree based on Hash is design for NAND [16]. The two-level index is used to manage file space, which can effectively reduce the time overhead of file search and creation. GlusterFS is a distributed user space file system without a special metadata server, and data and metadata are stored together [17]. In GlusterFS, there are two kinds of nodes such as client and server. There are no special nodes with special knowledge or responsibilities. So GlusterFS has no single point of failure. For most operations, GlusterFS is faster because it calculates location of metadata compared to retrieving metadata from any storage medium. It has high scalability and can handle thousands of clients at most. It divides the directories according to the number of files in the directory and file increments, and distributes large directories and small directories to different storage areas, which can reduce the time overhead of sequential access to directories by 17%, and reduce the GlusterFS rebalancing operation after adding servers 74% of the time overhead. A cluster of Metadata Server can be used to improve the efficiency of metadata management [18]. The directory subtree partitioning and the partition metadata consistent hash are designed to ensure cluster load balance and adapt to cluster change. In response to the unbalanced load, a dynamic load balance algorithm is proposed to adjust the load of the cluster at runtime.

3   Analysis of Metadata Management Mechanism

The metadata management is generally concentrated in the file system. They are organized by a directory tree and indexed by balanced tree or Hash. NVM has the characteristics of low latency, byte addressing, and fast reading and writing speed. There are some limitations for current metadata management mechanism with NVM.

Low search efficiency. All node should be accessed in path, when opening or accessing one file. The time overhead of traversing the directory tree is very expensive, especially when the directory tree is deep and contains a large number of files and directories. The location can be located quickly with the index based on Hash, but there are problems such as difficulty in supporting range queries and not supporting relative path access.

Bad concurrency. In order to avoid access conflicts of directory tree, all nodes in access path should be locked [19] when there is any modification of metadata. It will reduce the concurrency of metadata access seriously. Then it will affect the efficiency of management, especially in a computer with multi-core processor. More processor core will cause more serious access conflicts of metadata access. The index based on Hash also has the similar problem, its concurrency also needed to be improved and adapt to the multi-core system.

The Hash-based index needs a large calculation overhead. In order to reduce the conflict, the large Hash value and complex Hash function is need. Which will bring more time and space overhead. Especially with multi-core processor, concurrently metadata access will bring greater computational pressure to the CPU.

The metadata I/O software stack is expensive. The metadata needs to be read from the storage device and sent to the file system for processing, which requires multiple I/O software layers such as storage device drivers, page caches and file system. The long metadata I/O software stack will bring a lot of extra time overhead, and it is difficult to take advantage of the NVM device. At the same time, there are multiple NVM devices access for path traversal of metadata, which needs repeated transmission by the long metadata I/O software stack and affects the efficiency of metadata management.

The overhead of main memory is large. Cache or putting all metadata in main memory is an effective way to improve efficiency of metadata management, but the large capacity of main memory is needed. NVM devices already have a read and write speed close to main memory and it is an effective way to improve the efficiency of metadata storage and management.

In general, the current metadata management mechanism in file system is facing the pressure of increased access conflicts by the multi-core processor, the difficult to effectively take advantage of the fast read and write speed of NVM devices. Therefore, the new metadata management mechanism is very necessary to improve the efficiency and concurrency of metadata organization and management according to the characteristics of NVM devices and multi-core processor.

4   Structure of Two-Tier Hybrid Metadata Management Mechanism

We present the structure of a two-tier hybrid NVM metadata management mechanism, as shown in Fig. 1. The metadata management function of the existing file system is splitted into a hybrid distributed directory tree module and a multi-skiplists index embedded in driver module. The hybrid distributed directory tree module is responsible for decomposing the directory tree of file system. The first N layers of the directory tree are stored into DRAM to build a centralized metadata layer, and Hash is used to manage them. The metadata of the remaining levels are stored in the NVM device to build a dispersed metadata layer, and multi-skiplists is used to manage them. At the same time, the dual-zone cache is used to transfer the high-frequency metadata in the dispersed metadata layer into the cache located in the DRAM. The multi-skiplists index embedded in driver module uses multiple skiplists to store and manage a large amount of metadata in the dispersed metadata layer by the NVM device driver.

The two-tier hybrid NVM metadata management mechanism can reasonably decompose metadata according to the characteristics of metadata and the NVM device. The hot part of directory tree is stored in DRAM to ensure the efficiency, and the metadata decomposition can reduce the overhead of Hash value calculation and the

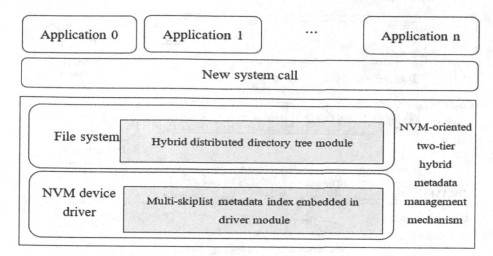

Fig. 1. The structure of two-tier hybrid metadata management mechanism.

Hash value conflict. At the same time, it will move part of the metadata management tasks to the NVM device driver, which can short the I/O system software stack for metadata. In addition, DRAM is used to build a cache for metadata in NVM devices to reduce the latency of metadata management.

5 Hybrid Distributed Directory Tree

B-tree is the popular to index the metadata. It is optimized for slow block storage devices such as HDD, which can reduce the access block from the HDD. But its concurrency is the problem for efficiency of computer with multi-core processor. Using Hash to index metadata also will meet the problem of high computational overhead and conflict rate, when the size of metadata is very large. Meanwhile, Hash also has the limitation of concurrency. Therefore, we design a hybrid distributed directory tree, as shown in Fig. 2. At first, the directory tree of the file system is decomposed, and the metadata is split into a centralized metadata layer and a dispersed metadata layer. The first N layers of the directory tree are the centralized metadata layer and stored in DRAM. The dispersed metadata layer contains all the metadata in the directory tree, which is stored in the NVM device. The Hash index is used for the centralized metadata layer and a new index based on multiple skiplists is used for the dispersed metadata layer. The metadata in the centralized metadata layer will be written back to the NVM device periodically or when necessary.

The hybrid distributed directory tree separates the top of the directory tree from the rest, they have the highest access frequency in generally. The DRAM is used to store them, which can take advantage of the faster read and write speed and higher concurrency of DRAM to ensure the efficiency [20]. At the same time, the capacity overhead of DRAM can be reduced, and the complexity and computational overhead of the Hash can also be reduced. The capacity advantage of NVM devices can be used to

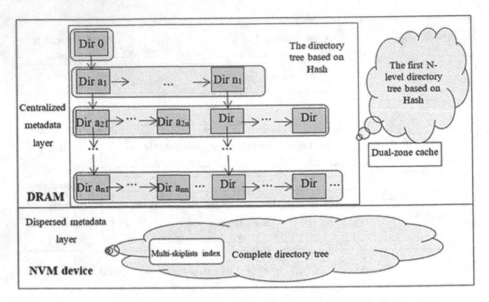

Fig. 2. The structure of the hybrid distributed directory tree.

store the complete directory tree and the multi-skiplists index can improve the concurrency of metadata management.

6 Dual-Zone Cache

Despite of the top item in the directory tree, some other parts also have a high access frequency during a certain period of time. Using DRAM to build a cache is a common method to improve access efficiency of these parts. Current cache management algorithms generally need to record access number, access time and other information. Meanwhile, these information is needed to be sorted for the cache candidate selection, which requires large additional time and space overhead. It will bring more burden for the metadata I/O software stack, so how to reduce the time overhead of cache management is an important issue to improve the efficiency of NVM storage system.

We design a dual-zone cache, its structure is shown in Fig. 3. At first, build a cache of the dispersed metadata layer in DRAM. The cache contains two parts: the cache area and the reserve area. Each entry in the cache area contains two parts: a cache item and an access count. The cache item is used to cache the metadata stored in the NVM device under the N level of the file directory tree with high access frequency. Each entry in the reserve area contains a reserve item address and an access count, and its access count is the times that the reserve item is accessed after it in the reserve area.

On this basis, we design a dual-zone cache management algorithm and the main steps of search metadata flow are given as follows:

Step 1: When accessing the metadata below the N level of the directory tree, the cache will be looked up firstly. If it is in cache area, the metadata will be feed back to the user, and access count of the corresponding cache item will be added 1. Then the look up flow will be end.

Step 2: Otherwise, searching the target in the reserve area, if it can be found and the current access count is not 0, the metadata will be read from NVM devices by the reserve item address and will be transmitted to the cache area. Then, the corresponding access count will be added 1 and assign it to the access count of new cache item. Then the look up flow will be end.

Step 3: If it can be found in the reserve area, but the expected value of the corresponding reserve item is 0, the corresponding access count will be added 1 and the metadata will be read by the reserve item address from NVM devices.

Step 4: If there is no corresponding item in the reserve area, the metadata will be read from the NVM device. The corresponding address will be recorded in the reserve area, the its access count will be set 1.

When the cache area or the reserve area is full, and new metadata should be added in the dual-zone cache, the main steps of the elimination flow are as follows:

Step 1: Looking for the item with zero access count in the cache area or the reserve area. If there is only one candidate, the cache item or the reserve item will be eliminated directly. If there are more than one candidate, the cache item with the longest entry time will be eliminated.

Step 2: If there are not the item with zero access count, the item in cache with the smallest access count is selected for elimination.

Meanwhile, the dual-zone cache adopts a refresh-per-access strategy. If there is a miss in the cache and the reserve area, all access count should be reduced by one.

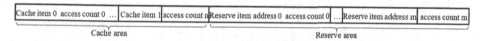

| Cache item 0 access count 0 | ... | Cache item 1 | access count n | Reserve item address 0 access count 0 | ... | Reserve item address m | access count m |

Cache area   Reserve area

Fig. 3. The structure of the dual-zone cache.

Using the dual-zone cache, DRAM can be used to build a cache for high-frequency access metadata in the dispersed metadata layer. Then, the directory tree can be decomposed dynamically, and distributed between DRAM and NVM devices. It can reduce the extra time and space overhead of cache management, and adapt to NVM devices with fast read and write speeds.

7 Multi-skiplists Index Embedded in Driver

Because NVM devices has fast read and write speed, low latency, and multiple concurrent channels inside, it is an important issue to improve the concurrency of the indexing mechanism of the dispersed metadata layer, especially in multi-core compute systems. Compared with index based on balanced tree or Hash, the skiplist has higher concurrency and has been widely used in many studies [21–24]. When the size of metadata is very large and many applications concurrent run in multi-core computer system, the concurrency is still a problem for index based on single skiplist. Based on the decomposed directory tree, we design an multi-skiplists index embedded in driver. One skiplist will be created for each lead of the directory tree in centralized metadata layer, and they will be embedded in the driver of NVM devices. Its structure is shown in Fig. 4. The six-tuple SL_Node (Node_parent_Id, Fs_name_hash, Dentry, Down, Next, Level) is used to represent the node of multi-skiplists. Node_parent_Id is the identity of the parent node. Fs_name_hash is the hash value of the access path without the first N layers. Dentry points to the directory of the file. Down points to the next level skiplist node. Next points to the next skiplist node. Level is the level number of this node.

The access path of metadata will be extracted and decomposed firstly, when the application accesses the metadata. If the layer of path is less than N, Hash will be used to calculate the address of metadata in the centralized metadata layer. Otherwise, the top N layers in access path will be used to calculate the Hash value, and search in the centralized metadata layer to obtain the entry address of the corresponding metadata skiplist in the dispersed metadata layer. Then, the special skiplist will be used to search the metadata by the other level of the access path.

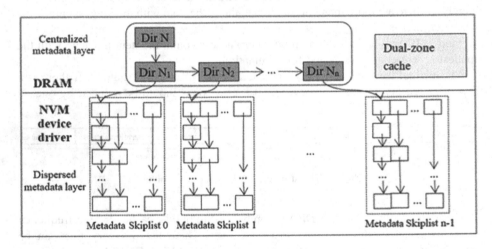

Fig. 4. Multi-skiplists index embedded in driver.

Therefore, the multi-skiplists index embedded in driver can improve the concurrency of metadata management stored in NVM devices, which can adapt to the high access conflicts in a multi-core computer system. The number of nodes in each metadata skiplist and the conflict rate of a single metadata skiplist can also be reduced to improve the metadata management efficiency. Meanwhile, the index is embedded in the NVM device driver, which can simplify the I/O system software stack for metadata management and avoid repeated transmissions in several software layers. Thereby, it can improve the efficiency of metadata management based on NVM devices.

8 Prototype and Analysis

We give the implementation of the prototype HSNVMMS and use general test tools for analysis.

8.1 Prototype and Test Environment

At first, we modified the source code of the Intel Optane DC Persistent Memory device driver named PMEM, the EXT4 and the Linux-4.4.112 kernel to implement the prototype of the two-tier hybrid metadata management for NVM storage system (HSNVMMS). The directory tree in EXT4 is decomposed and split into two parts: a centralized metadata layer and a dispersed metadata layer. The centralized metadata layer contains the metadata of the first two layers in the directory tree, stored in DRAM, and managed by EXT4 using hash-based method for management. The rest metadata of the directory tree belongs to the dispersed metadata layer, and are managed by PMEM using the multi-skiplists index. At the same time, the dual-zone cache is used to replace the page cache.

For comparison, the EXT4 and NOVAs are loaded on Intel Optane DC Persistent Memory, and implement two comparative prototypes named EXT4-NVM and NOVA-NVM. The Webserver, Varmail and Fileserver workload in Filebench are used to simulate typical application system workload for testing and analysis.

One server with the Intel Optane DC Persistent Memory is used to build the test environment for three prototypes, and the configuration is shown in Table 1.

Table 1. Configuration of the test environment

Part	Configuration
CPU	Intel Xeon CPU E5–2620 v3 @ 2.40 GHz * 2
Memory	128 GB
NVDIMM	2 * 128 GB Intel Optane DC Persistent Memory
Disk	600 GB SAS Disk
OS	CentOS 7.0 with Kernel 4.4.112

8.2 Webserver

At first, the Filebench with the Webserver workload is used to test the throughput of three prototypes. The ratio of read and write access requests for the Webserver workload is 10:1, the number of test threads is set to 1, the size of each I/O is 4 KB, and the test time is 60 s. The number of files is set to 10000, 50000, 100000, 200000, 300000 and 500000 respectively. The results are shown in Fig. 5.

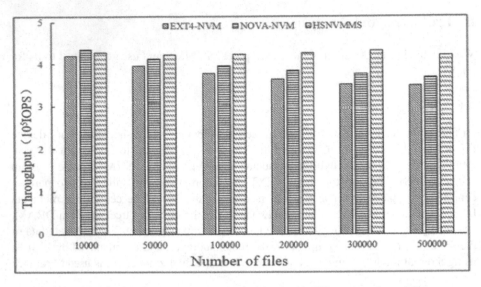

Fig. 5. Test results under Webserver workload with different number of files.

The results in Fig. 5 show that HSNVMMS can effectively increase the IOPS under Webserver workload. Compared with EXT4-NVM and NOVA-NVM, it can increase IOPS by 2% ~ 22.6% and 14.6%, respectively. As the number of files is increased from 10000 to 500000, the IOPS of EXT4-NVM and NOVA-NVM keeps going down, and decreases by 16.96% and 15.2% respectively. This is because more files will increase conflicts between metadata access requests. This leads to a drop in IOPS. Using HSNVMMS, when the number of files is 10000, the IOPS is higher than EXT4-NVM and lower than NOVA-NVM, but when the number of files increases to 50000, the IOPS of HSNVMMS is higher than EXT4-NVM and NOVA-NVM. At the same time, as the number of files increases to 300,000, the IOPS of HSNVMMS keeps growing. These results show that the hybrid distributed directory tree and the multi-skiplists index embedded in driver can reduce metadata access conflicts in HSNVMMS to improve the throughput, which can better adapt to computer with multi-core processor and improve metadata concurrent access performance.

Using the Webserver workload and the above parameters, but the number of threads is set to 1, 4, and 8 respectively. The results are shown in Fig. 6.

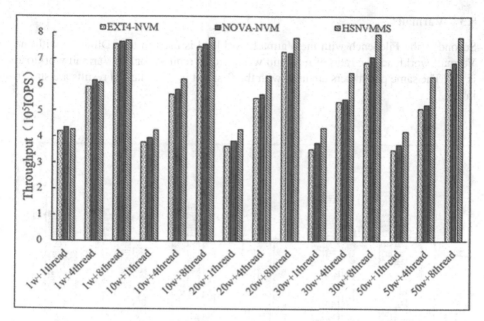

Fig. 6. Test results under Webserver workload with different number of files and threads.

As seen from Fig. 6, HSNVMMS can improve the IOPS more significantly with multiple threads compared with EXT4-NVM and NOVA-NVM. When the number of files is 10,000, HSNVMMS is slightly lower than NOVA-NVM with 1 and 4 threads, but it has higher IOPS than NOVA-NVM with 8 threads, and its IOPS is higher than NOVA-NVM under other conditions. In addition, HSNVMMS has higher IOPS than EXT4-NVM in all test cases. With the increase of the number of threads and files, the IOPS improved by HSNVMMS compared to EXT4-NVM and NOVA-NVM is still increasing, from only 2% and 0.7% with 10000 files and 8 threads, to 23.9% and 20.9% with 500000 files and 4 threads. These results fully show that the hybrid distributed directory tree and the multi-skiplists index embedded in driver in HSNVMMS can adapt to the high conflict characteristics of the computer with multi-core processor, and effectively improve the efficiency of metadata management. At the same time, when the number of files is less than 300000, the ratio of IOPS increased by HSNVMMS compared to EXT4-NVM and NOVA-NVM will gradually decrease with the increase of thread number. When the number of files is more than 300000, after increasing the number of threads, the ratio of IOPS improvement by HSNVMMS will increase with the rise of thread number and reach the highest value at 4 threads. It shows that HSNVMMS is suitable for improving the management efficiency with a large amount of metadata.

8.3   Varmail

Secondly, the Filebench with the Varmail workload is used to test. Different with the Varmail workload, the ratio of read and write access request for the Varmail workload is 1:1. The same parameters are used with the first test in 8.2. The test results are shown in Fig. 7.

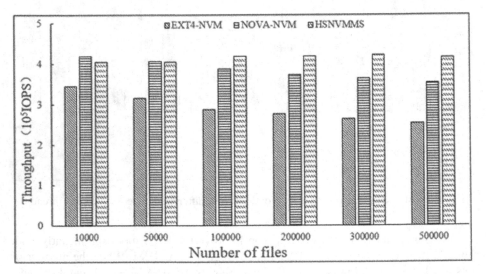

Fig. 7. Test results under the Varmail workload with different number of files.

It can be seen from Fig. 7 that the IOPS of HSNVMMS can improve higher under the Varmail workload than the Webserver workload. Compared with EXT4-NVM and NOVA-NVM, the IOPS of HSNVMMS increased by 17.1% ∼ 59.9% and 15.9%. When the number of files increases from 10000 to 500000, the IOPS of EXT4-NVM and NOVA-NVM decreased by 26.6% and 15.8%, respectively. On the contrary, HSNVMMS can still increase IOPS by 4.5% when the number of files grows from 10000 to 300000, and remains basically stable when the number of files grows to 500000. This is due to a more balanced ratio of read and write access request in the Varmail workload, which improves metadata access conflicts and reduces the IOPS of EXT4-NVM and NOVA-NVM. At the same time, these results also further verify that HSNVMMS can improve the concurrency of metadata access, short the I/O system software stack, and improve the efficiency of metadata management in computer multi-core processor with more access request conflicts.

Using the Varmail workload and the same parameters of the second test in 8.2 to test the IOPS of three prototypes respectively. The results are shown in Fig. 8.

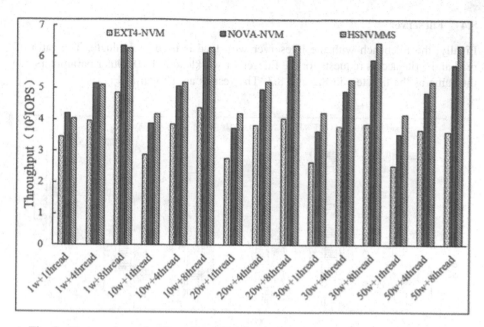

Fig. 8. Test results under Varmail workload with different number of files and threads.

It can be found from Fig. 8 that the overall trend is similar to those under the Webserver workload. However, due to the high percentage of write access requests in the Varmail workload, the percentage of IOPS improvement of HSNVMMS compared with NOVA-NVM has decreased. Compared with EXT4-NVM, the ratio of IOPS that can be improved is greatly improved and reaches 17.1%−76.9%. At the same time, the IOPS of HSNVMMS is always higher than EXT4-NVM and NOVA-NVM except when the number of files is 10000. It is different with under the Webserver workload, except that the number of files is 10000, the ratio of IOPS that can be increased by HSNVMMS with 4 threads is lower than that of single thread and 8 threads. When the number of files is unchanged and the number of threads is increased separately, the IOPS of HSNVMMS and NOVA-NVM will both increase, but the IOPS of EXT4-NVM will decrease when the number of threads increases from 4 to 8 with 500000 files. At the same time, when the number of threads increases from 1 to 8, the increase rate of HSNVMMS's IOPS has little to do with the number of read and write files, and it stabilizes at about 23%. However, the IOPS improvement ratio of EXT4-NVM and NOVA-NVM will decrease sharply as the number of threads increasing and even only about 2%. These results show that HSNVMMS can better adapt to multi-threaded applications and utilize the advantages of multi-core processors than EXT4-NVM and NOVA-NVM, and improve the efficiency of metadata management.

8.4   Fileserver

Finally, the Filebench with the Fileserver workload is used to evaluate. The ratio of read and write access requests for the Fileserver workload is 1:10. Other parameters are the same as the first test in 8.2 and 8.3. The results are shown in Fig. 9.

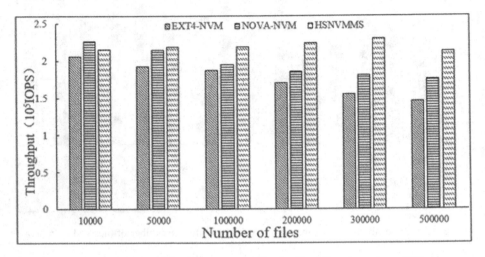

Fig. 9. Test results under Fileserver workload with different number of files.

It can be seen from Fig. 9 that the IOPS has the similar trend under the Fileserver workload compared with above workloads. HSNVMMS can increase the IOPS by 4.7% ~ 48.3% and 26.9% compared with EXT4-NVM and NOVA-NVM respectively. As the number of files increases, the IOPS of EXT4-NVM and NOVA-NVM continued to decline, down 29.4% and 22.4%, respectively, which is the highest drop rate among the three loads. Meanwhile, the IOPS under the Fileserver workload is the lowest among the three workloads. This is because the Fileserver workload has the highest proportion of write access requests among the three loads. A large number of write access requests will bring more metadata access conflicts and reduce the efficiency of file system metadata management. When the number of files increases from 10000 to 300000, the IOPS of HSNVMMS keeps growing and the increase ratio is 6.4%. These results further show that HSNVMMS can effectively improve the concurrency of metadata management.

Then the Fileserver workload is used to test the IOPS with the same parameters as the second test in 8.3 and 8.4. The results are shown in Fig. 10.

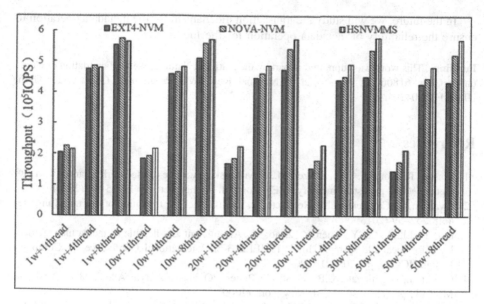

Fig. 10. Test results under Fileserver workload with different number of files and threads.

It can be found from Fig. 10 that trend under the Fileserver workload is similar to it under the Webserver workload and the Varmail workload. The IOPS of HSNVMMS is always higher than that of EXT4-NVM, and it is higher than that of NOVA-NVM except when the number of files is 10000. Increasing the number of threads and files can effectively improve the IOPS of HSNVMMS. Compared with under the Varmail workload, HSNVMMS can improve the ratio of IOPS higher than NOVA-NVM. This is because the Fileserver workload contains more write access requests, which further reflects the advantages of HSNVMMS. However, the increased IOPS ratio by HSNVMMS is greatly reduced compared with EXT4-NVM. In addition, the change of increased IOPS ratio is larger than under the Webserver workload. This is also due to the fact that there are more write access requests in the Fileserver workload, which results in a higher access conflict rate and raises the demand for concurrency of metadata management.

9   Conclusion

We analyze the characteristics of NVM devices and the current metadata management methods. Then, we designed the two tier hybrid metadata management mechanism for NVM storage system. A hybrid distributed directory tree and multi-skiplists index embedded in driver are given to improve the efficiency and concurrency of metadata management through metadata decomposition, efficient caching and high-concurrency indexing. Three typical application workloads are used to evaluate and the results show the effectiveness of the two tier hybrid metadata management mechanism.

In the future, we will study the transaction mechanism for the metadata operation to ensure the reliability of metadata operation for the file system.

Funding. This work was supported in part by the National Natural Science Foundation of China under Grant 61806086, the Project of National Key R&D Program of China under Grant 2018YFB0804204.

References

1. Mason, T., Doudali, T.D., Seltzer, M., Gavrilovska, A.: Unexpected performance of Intel® Optane™ DC persistent memory. IEEE Comput. Archit. Lett. **19**(1), 55–58 (2020)
2. Swanson, S., Caulfield, A.M.: Restructuring the I/O stack for the future of storage. Computer **46**(8), 52–59 (2013)
3. Liu, Y., Li, H., Lu, Y., Chen, Z., Zhao, M.: An efficient and flexible metadata management layer for local file systems. In: 2019 IEEE 37th International Conference on Computer Design (ICCD), pp. 208–216 (2019)
4. Condit, J., Nightingale, E.B., Frost, C.: Better I/O Through Byte Addressable, Persistent Memory, pp. 133–146. ACM, New York (2009)
5. Wu, X., Reddy, A.L.N.: SCMFS: a file system for storage class memory. In: SC 11: Proceedings of 2011 International Conference for High Performance Computing, Networking, Storage and Analysis, pp. 1–11 (2011)
6. Ipek, E., et al.: System software for persistent memory. In: Proceeding of the 9th European Conference on Computer System, pp. 1–15. ACM, New York (2014)
7. Xu, J., Swanson, S.: NOVA-NVM: a log-structured file system for hybrid volatile/non-volatile main memory. In: Proceeding of the 14th Conference on File and Storage Technologies, pp. 323–338. USENIX, Berkeley, CA (2016)
8. Xu, J., Zhang, L., Memaripour, A.: NOVA-Fortis: a fault-tolerant non-volatile main memory file system. In: Proceeding of the 26th Symposium on Operating System Principles, pp. 478–496. ACM, New York (2017)
9. Tao, C., Niu, D., Yao, H., Zhu, Y.: "NVMCFS: complex file system for hybrid NVM. In: 2016 IEEE 22nd International Conference on Parallel and Distributed Systems, pp. 577–584. IEEE, Wuhan (2016)
10. Jin, P.Q., Wu, Z.L., Wang, X.L.A.: Page-based storage framework for phase change memory. In: Proceedings of the 2017 International Conference on Massive Storage Systems and Technology, pp. 152–164. IEEE, Piscataway (2017)
11. Fan, Z.Q., Wu, F.G., Park, D.: Hibachi: a cooperative hybrid cache with NVRAM and DRAM for storage arrays. In: Proceeding of the33rd International Conference Mass Storage Systems and Technologies, pp. 26. IEEE, Piscataway, NJ (2017)
12. Li, S., Huang, L.: LosPem: a novel log-structured framework for persistent memory: In: ACM Journal on Emerging Technologies in Computing Systems, pp. 1–17 (2020)
13. Kwon, Y., Won, Y., Fingler, H., Hunt, T.: Strata: a cross media file system. In: Proceedings of the 26th ACM Symposium on Operating Systems Principles, pp. 460–477. ACM, New York (2017)
14. Shengan, Z., Morteza, H., Steven, S.: Ziggurat: a tiered file system for non-volatile main memories and disks. In: Proceeding of the 17th Conference on File and Storage Technologies, pp. 207–219. USENIX Association, Berkeley, CA (2019)
15. Chou, C.C., Jung, J., Reddy, A.L.: Virtualize and share non-volatile memories in user space. In: CCF Transaction on High Performance Computing, pp. 16–35 (2020)

16. Lim, S., Lee, C., Park, K.: Hashing directory scheme for NAND flash file system. In: The 9th International Conference on Advanced Communication Technology, pp. 273–276 (2007)

17. Fattahi, T., Azmi, R.: A new approach for directory management in GlusterFS. In: 2017 9th International Conference on Information and Knowledge Technology (IKT), pp. 166–174 (2017)

18. Huo, L., Yi, R.: Research on metadata management scheme of distributed file system. In: 2015 International Conference on Computer Science and Applications (CSA), pp. 37–41 (2015)

19. Cederman, D., Gidenstam, A., Ha, P.H., Sundell, H., Papatriantafilou, M., Tsigas, P.: Lock-free concurrent data structures. In: Pllana, S., Xhafa, F. (eds.) Programming Multi-Core and Many-Core Computing Systems. Wiley-Blackwell, New Jersey (2014)

20. Salkhordeh, R., Mutlu, O., Asadi, H.: An analytical model for performance and lifetime estimation of hybrid DRAM-NVM main memories. IEEE Trans. Comput. **68**(8), 1114–1130 (2014)

21. Pugh, W.: Skip lists: a probabilistic alternative to balanced trees. Commun. ACM **33**(6), 668–676 (1990)

22. Shavit, N., Lotan, I.: Skiplist-based concurrent priority queues. In: Proceedings 14th International Parallel and Distributed Processing Symposium, pp. 263–268 (2000)

23. Chen, Q., Yeom, H.: Design of skiplist based key-value store on non-volatile memory. In: 2018 IEEE 3rd International Workshops on Foundations and Applications of Self* Systems (FAS*W), pp. 44–50 (2018)

24. Renzhi, X., et al.: Write-optimized and consistent skiplists for non-volatile memory. IEEE Access **9**, 69850–69859 (2021)

A Novel CFLRU-Based Cache Management Approach for NAND-Based SSDs

Haodong Lin[1], Jun Li[1], Zhibing Sha[1], Zhigang Cai[1(✉)], Jianwei Liao[1], and Yuanquan Shi[2(✉)]

[1] College of Computer and Information Science, Southwest University, Chongqing, China
czg@swu.edu.cn
[2] School of Computer Science and Engineering, Huaihua University, Huaihua, China

Abstract. To ensure better I/O performance of NAND-based SSD storage, a DRAM cache is commonly equipped inside SSDs to absorb overwrites or writes performed to underlying SSD cells. This paper presents a simple, effective cache management method based on clean first least recently used (*CFLRU*) for SSDs, by also taking the factor of spatial locality into account. That is, we introduce a novel data management approach to separate hot and cold buffered data in the SSD cache by considering both factors of temporal and spatial locality, when accessing or inserting a piece of write data in the SSD cache. As a result, the hot buffered data and their spatially adjacent buffered data can be preferentially kept in the cache and other cold data will be evicted first. Simulation tests on several realistic disk traces show that our proposal improves cache hits by up to 4.5%, and then cuts down I/O response time by up to 9.4%, in contrast to the commonly used *CFLRU* scheme.

1 Introduction

The NAND-based solid-state drivers (SSDs) have been deployed in a wide range of digital devices, data centers and supercomputers, thanks to their advantages of random access, lower power consumption, and collectively massive parallelism [1–3]. Apart from a NAND flash array that holds data, an SSD device generally has a faster but small scale Dynamic Random Access Memory (DRAM) as the cache inside SSD. For example, the Cosmos OpenSSD platform [4], which is publicly released by the OpenSSD project, is equipped with more than 100 MB memory. Generally, the SSD memory is used to not only keep address mapping data structures, but also temporarily buffer the contents of overwrite or write requests [5,6]. Consequently, it can cut down the number of flush operations onto the underlying flash array and then improve I/O performance of SSD devices [7].

© IFIP International Federation for Information Processing 2022
Published by Springer Nature Switzerland AG 2022
C. Cérin et al. (Eds.): NPC 2021, LNCS 13152, pp. 214–225, 2022.
https://doi.org/10.1007/978-3-030-93571-9_17

In other words, the SSD cache improves performance by keeping recent or often-updated data items (i.e. data pages) in RAM locations that are faster than flash memory stores [5]. When the cache is full, the cache management method must choose which victim items (i.e. data pages) to be flushed out and to make room for the new ones [8,9]. As a rule, the cache management schemes in the context of SSDs can be classified into the following two categories, according to their internals and application scenarios.

To be specific, the first type is *simple and universal*: least recently used (LRU) is the most common scheme because of the simplicity and reasonable performance gains [10]. Clean first least recently used ($CFLRU$) is another representative cache management in SSDs, which is an enhancement on the basis of the LRU scheme [11]. $CFLRU$ maintains a list corresponding to the logic page numbers of cached data pages. The head node of list is the most recently used page and the tail node is the least recently used page. Difference from LRU, the $CFLRU$ scheme looks forward from the tail of the linked list for the buffered data pages that have not been modified in the cache, and treats them as eviction victims. This is because evicting the modified data pages indicates flushing the pages onto the flash array of SSD, which must result in more time overhead [11].

Another type is *specialized and more efficient*: Sun et al. [9] proposed a collaborative active write-back cache management scheme, which is collaboratively aware of I/O access patterns and the idle status of flash chips. Thus, it can minimize the negative effects of cache eviction. Similarly, Chen et al. [8] introduced ECR, that gives a higher probability to evict a page when it causes the shortest waiting time in the corresponding chip queue. Besides, Wang et al. [5] provided a scheme for the management of SSD cache with consideration of the access frequency of the buffer pages. Specially, they used the particle swarm optimization (PSO) technique [12] to make predictions of access frequency of the pages, for directing cache replacement.

The relevant verification experiments show that the specialized cache management approaches can greatly enhance cache use efficiency in many cases by comparing with the simple and universal cache management schemes. But, such complicated methods consume significant computational power due to analyzing I/O workloads or monitoring the idle status of SSD devices, and cannot cover all scenarios with expected I/O improvements [5]. Considering SSD devices normally have computing power-limited controllers, it is expected to integrate a simple and effective cache management scheme with such devices.

On the other side, we understand that the factors of spatial locality and temporal locality of workloads are the root causes for cache designs to boost the storage performance in computing systems [13]. The spatial locality of reference, however, has not been addressed together with the factor of temporal locality in commonly used cache management schemes in SSDs.

To address the aforementioned issues, this paper intends to offer a cache management scheme for SSD devices on the basis of $CFLRU$ (called as *Batched-CFLRU*), by also taking the factor of spatial locality into account. In brief, it has the following two contributions:

- We introduce a cache data management approach to separate hot and cold buffered data, by unifying the spatial and temporal locality, when accessing or inserting a piece of write data in the SSD cache. It employs three levels of linked lists, for upgrading the hot buffered data and their neighboring buffered data to high-level linked lists *in batch* and degrading the cold data to low-level linked lists till the data are evicted from the SSD cache.
- We offer preliminary evaluation on several block I/O traces of real-world applications. As our measurements show, the proposed method can noticeably increase the cache hits and then reduce the I/O time by 3.1% on average, in contrast to commonly used approach of *CFLRU*.

The rest of paper is organized as follows: Sect. 2 depicts related work and motivations. We will demonstrate design details of the proposed cache management scheme by referring to both temporal and spatial locality in Sect. 3. The evaluation methodology, experimental results and relevant discussions are illustrated in Sect. 4. Finally, we conclude the paper in Sect. 5.

2 Related Work and Motivations

2.1 Related Work

Caching inside of SSD can absorb certain overwrite and write requests to optimize SSD performance. A number of prominent studies were conducted to shape cache management, that mainly focus on the replacement strategy to make room for the new data by ejecting the buffered data in the SSD cache. The universal cache replacement strategies are built on the temporal locality of reference, and include first in first out (*FIFO*), *LRU*, *CFLRU*, and least frequently used (*LFU*), among which *LRU* and its variation of *CFLRU* are the most commonly employed schemes [9,11].

More specifically, Park et al. [11] proposed the *CFLRU* cache replacement policy for NAND-based flash memory. It holds two LRU lists including a working list and a clean-first list, and prefers to evict clean pages from the cache if they are not read frequently. Thus, *CFLRU* works efficiently for universal cases, and can greatly reduce the number of costly flush operations onto underlying flash cells and potential erase operations, though it is simple.

Furthermore, many sophisticated cache management methods have been proposed to address cache management for certain specific application scenarios. Wang et al. [5] presented to consider the access frequency of the buffer pages for directing cache ejection. To this end, they make use of the advanced particle swarm optimization (PSO) algorithm [12] to make predictions of access frequency of the pages. After that, the (predicted) less frequently accessed pages will be preferentially evicted. In order to eliminate unnecessary SSD cache writes and improve cache space utilization, Wang et al. [14] introduced a machine learning approach by integrating a decision tree into the classifier, to predict which data should be directly flushed onto flash array, instead of keeping them in the cache. Yoon et al. [15] proposed a cache replacement policy for SSDs, by analyzing the access characteristic of workloads, to reduce the number of cache evictions due to partial-page updates.

Besides, Wu et al. [16] presented a garbage collection aware replacement policy (called as *GCaR*). Different from the traditional cache management schemes, *GCaR* not only considers the temporal locality of workloads, but also makes the miss penalty as an important factor in a cache replacement process. Then, it will assign a higher priority to cache the data blocks belonging to the flash chips that are currently enduring a GC operation. Similarly, Chen et al. [8] proposed eviction-cost-aware cache management policy (called *ECR*), that gives a higher priority to keep a page in the cache when it will have a long waiting time in the corresponding chip queue.

The sophisticated cache management methods cost computing power for analyzing I/O workloads or collecting SSD idle status, to make an "accurate" ejection decision. Then, we emphasize that such approaches are not good solutions for computing power-limited SSD devices, and it is necessary to propose a simple but more effective cache management scheme on the basis of the widely used (simple) scheme of *CFLRU*.

2.2   Motivation

As discussed, existing cache management approaches for SSDs, such as *LRU*, *CFLRU*, and *LFU*, only take the factor of temporal locality into account when carrying out data replacement. In order to verify whether the factor of spatial locality matters or not in cache management of SSDs, we have conducted a series of trace-driven simulation experiments and collected relevant results. Read Sect. 4.1 about the specifications on experimental settings including the platform and benchmarks.

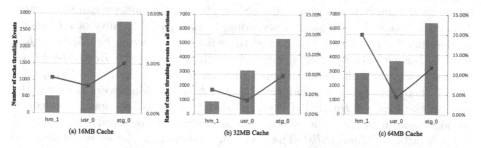

Fig. 1. The number of cache thrashing events and the ratio of such events to the total number of evictions (with *CFLRU*), with varied cache configurations.

To this end, we first define a term of **cache thrashing**, and a cache thrashing event indicates the cached data item is evicted out and loaded into the cache again **meanwhile** its (address) neighboring data items are invariably kept in the SSD cache. After that, we record the number of cache thrashing events and count the ratio of such events to the total evicted data items.

Figure 1 shows the statistic results of cache thrashing when using the commonly used cache management scheme of *CFLRU*. Clearly, there are a considerable quantity of cache thrashing events, and such events take a non-negligible part of total evicted data pages after running the selected benchmarks. That is to say, it is possible to avoid a cache thrashing case if the thrashed data item can be managed in a batch with its in-cached neighboring data items. Thus, we conclude that the spatial locality of cached pages is worth considering in cache management of SSDs, to boost cache use efficiency and I/O performance.

Observation: *Meanwhile many cached data pages are evicted from the cache and loaded into the cache again in a short interval, their neighboring data pages are invariably buffered in the cache during this period.*

Such observation motivates us to introduce an efficient and simple approach based on *CFLRU* for cache management inside SSDs, by unifying both temporal and spatial locality of references. Consequently, the I/O performance of SSDs can be improved for running a wide range of applications.

3   Batched-CFLRU Cache Management Scheme

3.1   High-Level Overview of *Batched-CFLRU*

The basic principle of our approach (termed as *Batched-CFLRU*) is to take both spatial locality and temporal locality into account, for directing cache management in SSDs. To this end, we define three-level lists helping the management of cached data pages, and each node in the lists corresponds to a buffered data page. Then, it can sift the hot buffered data pages and their neighboring data pages, to preferably keep them in the SSD cache. As a result, we can yield gains if they are accessed again shortly according to the spatial locality of reference.

Specifically, our proposal maintains three lists, i.e. *Eviction list*, *Monitor list*, and *Hot list* with the ascending order. Basically, the proposed *Batched-CFLRU* method evicts the buffered data pages to make space for the new data, *if-and-only-if* their corresponding nodes are in *Eviction list*. Moreover, it moves the node to the head of *Hot list* if the cached data page is hit again, and correspondingly adjusts the nodes of its spatially adjacent (buffered) data page(s) to the head(s) of *Monitor list*. That is to say, we manage the accessed buffered data pages and their neighboring buffered data pages in the unit of batch. As a result, the hot accessed data and their neighboring buffered data can be preferably buffered in the SSD cache for possibly enduring the future I/O requests that follows the locality of references.

3.2   Batch Adjustment of Cached Pages

This section depicts the specifications of batch adjustment on nodes in three-level lists of *Batched-CFLRU*, which correspond to the buffered data pages.

Algorithm 1: Batch Adjustment in Cache Management

Input: The I/O request of REQ_{lsn}
Output: $NULL$

1 **Function** $main\_routine()$
2   /*the read request on the page of $lsn(logic\ sector\ number)$*/
3   **if** $is\_a\_read(REQ_{lsn})$ **then**
4   **if** $is\_incache(lsn)$ **then**
5           $read\_buffered\_data(lsn)$;//read data page from cache
6           $move\_to\_list\ (Hot\_list,\ lsn)$;
7   **else**
8           $read\_from\_flash(lsn)$;//read data page from underlying flash

9   /*the write request on the page of lsn*/
10   **else**
11   /*hit again in the cache, adjust $Hot\ list$*/
12   **if** $is\_incache(lsn)$ **then**
13          $update\_buffered\_data(lsn)$;
14          $move\_to\_list\ (Hot\_list,\ lsn)$;
15   **else**
16   /*evict a buffered data pages for making cache space*/
17   **if** $is\_cache\_full()$ **then**
18   **if** $is\_eviction\_list\_empty()$ **then**
19                $degrade\_lists()$;
20             $eject\_from\_list\ (Eviction\_list)$;
21   /*insert a new node into Hot_list*/
22          $insert\_to\_list\ (Hot\_list,\ lsn)$;

23   **if** $is\_inlist(Eviction\_list,\ lsn\pm1)$ or $is\_inlist(Monitor\_list,\ lsn\pm1)$ **then**
24   /*adjust neighboring buffered data pages to head(s) of $Monitor\ list$ */
25       $move\_to\_list\ (Monitor\_list,\ lsn\pm1))$;

At the initialization stage (i.e. all lists are empty), all the nodes of cached data pages are linked in the *Eviction list* by following the manner of *CFLRU*. Once a given cached page is hit again, *Batched-CFLRU* carries out the upgraded batch adjustment to keep the nodes of hot hit page and their spatially adjacent (cached) pages in the higher-level lists.

On the other side, *Batched-CFLRU* requires to conduct hierarchical downgraded movements when a lower-level list becomes empty. That is, if the last node of *Eviction list* is removed and *Eviction list* becomes empty. Thus, all nodes of *Monitor list* and *Hot list* should be degraded as the elements of *Eviction list* and *Monitor list* respectively.

3.3 Implementation Specifications

Algorithm 1 demonstrates the specifications on batch adjustment on the nodes in three-level linked lists, that are representations of buffered data pages in cache management of SSDs.

As seen, *Lines 3–8* aim to service a read request with the cached data or the data obtained from underlying flash cells. *Lines 10–22* deal with a write request. Specially, it has to make cache space for the new write data, by evicting a buffered data page whose list node is the tail of *Eviction list*, when the cache is full. Finally, if the requested data page is hit in the cache or a new write data page is loaded into the cache, it triggers a batch adjustment in the relevant lists, as shown in *Lines 23–25*.

4 Experiments and Evaluation

This section first describes the experimental settings. Then, the evaluation results and relevant discussions are presented to show the effectiveness of our proposal. At last, we make a summary of the findings gained from the tests.

4.1 Experimental Setup

Table 1. Experimental settings of *SSDsim*

Parameters	Values	Parameters	Values
Channel size	16	*Read latency*	0.075 ms
Chip size	2	*Write latency*	2 ms
Plane size	2	*Erase latency*	15 ms
Block per plane	810	*Transfer (Byte)*	10 ns
Page per block	384	*GC threshold*	10%
Page size	8 KB	*DRAM cache*	16/32/64 MB
FTL scheme	Page level		

Table 2. Specifications on selected disk traces (ordered by the write ratio)

Traces	Req #	Wr ratio	Wr size	Frequent R (Wr)
web1	1050001	0.02%	8.6 KB	94.1% (0.009%)
hm_1	609310	4.7%	20.1 KB	46.6% (37.4%)
usr_0	2237968	59.6%	10.3 KB	50.9% (57.4%)
wdev_0	1143260	79.9%	8.2 KB	69.3% (82.2%)
ts_0	1801722	82.4%	8.0 KB	45.4% (74.2%)
stg_0	2030915	84.8%	9.2 KB	73.4% (90.9%)

Note: **Frequent R** means the ratio of addresses requested not less than 3, and **(Wr)** implies the percent of write addresses in which.

We have performed trace-driven simulation with *SSDsim (ver2.1)* [17], which has been modified to support the newly proposed cache management scheme, on a local ARM-based machine. The machine contains an ARM Cortex A7 Dual-Core with 800 MHz, 128 MB of memory and runs 32-bit Linux (*ver*3.1). Table 1 demonstrates our settings of experiments, by mainly referring to [18,19]. Specially, in order to further investigate how our proposal works with different scales of the SSD cache, we set the cache size varying from 16 MB to 64 MB.

In addition, we employed 6 commonly used disk traces. Specifically, *web-search_1* (label as *web1*) is collected from UMass Trace Repository [20], and the next five traces are from the block I/O trace collection of Microsoft Research Cambridge [21]. The detailed specifications on the traces are reported in Table 2. Then, we show the effectiveness of the proposed *Batched-CFLRU* method, by comparing it with the commonly used *CFLRU* method, after replaying the selected traces.

4.2 Results and Discussions

Cache Hits and Thrashing. We first define the metric of the number of cache hits without flushing the buffered data onto underlying SSD cells. This term means the write data can be directly saved in the cache without ejecting other buffered data to make room for the new data. In other words, if the write requests can be directly absorbed in the cache, the corresponding write requests can be completed with less latency.

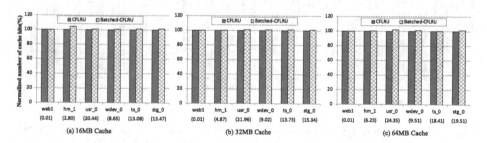

Fig. 2. Comparison of cache hits with varied cache configurations (the number underlying the traces are the absolute values with *CFLRU*, *unit:* 10^6).

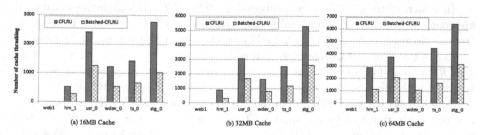

Fig. 3. Comparison of cache thrashing with varied cache configurations.

Figure 2 reports the results of cache hits after running the benchmarks with *CFLRU* and *Batched-CFLRU*. As illustrated, *Batched-CFLRU* does achieve an improvement on cache hits by up to 4.5% comparing with *CFLRU*. Note that *Batched-CFLRU* does not noticeably outperform *CFLRU* when running *web1* and *hm_1*. We argue both traces are read-intensive and a considerable part of write data can be buffered in the SSD cache, for example, all of write data of *web1* can be hold in the cache. Thus, cache management schemes do not make many differences.

To further disclose the cause of the increase of cache hits resulted by our proposal, we have recorded the number of cache thrashing events and Fig. 3 shows the results. Clearly, *Batched-CFLRU* can reduce the number of cache thrashing events by 53.2% on average, in contrast to *CFLRU*. This is because keeping the neighboring data pages of currently accessed data pages in the SSD cache with batches can better ensure the spatial locality of reference, which does contribute to the reduction of cache thrashing and thus improves the measure of cache hits.

I/O Latency. Considering the I/O response time greatly varies from case to case, we then count the normalized I/O response time of all selected traces, and Fig. 4 presents the results.

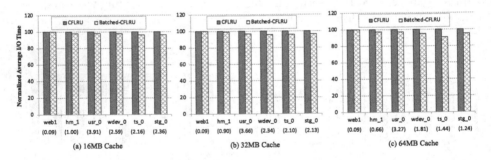

Fig. 4. Comparison of I/O response time with varied cache configurations (the number underlying the traces are the absolute values with *CFLRU*, **unit: ms**).

In contrast to *CFLRU*, the proposed *Batched-CFLRU* cache management approach brings about a noticeable improvement on I/O response time in all traces. More exactly, *Batched-CFLRU* can cut down the average I/O time by up to 9.4% and by 3.1% on average. More importantly, we can see that *Batched-CFLRU* can yield more obvious I/O improvements when the cache size becomes larger, which demonstrates our proposal does have well scalability.

Another noticeable clue is about the proposed *Batched-CFLRU* scheme does not distinctly perform better than *CFLRU*, when replaying two read-intensive workloads of *web1* and *hm_1*. We argue that both *web1* and *hm_1* have a very small number of write requests (i.e. 212 and 28223), that limits the room for

I/O improvements with the cache. On the other side, *Batched-CFLRU* can bring about more than 4.6% reduction of I/O response time for write-intensive workloads of *usr_0*, *wdev_0*, *ts_0*, and *stg_0*. This fact verifies *Batched-CFLRU* offers better cache use efficiency (i.e. more cache hits), which greatly contribute to the reduction in I/O response time for write-intensive workloads.

Analysis on Batch Adjustment. Different from *CFLRU* that only adjusts the accessed data page in the cache, our proposal may adjust multiple data pages in batch, with respect to processing an I/O request. Then, we record the number of batch adjustment and the ratio of batch adjustment operations to all I/O requests, after replaying the selected traces with *Cached-CFLRU*. Figure 5 reports the results.

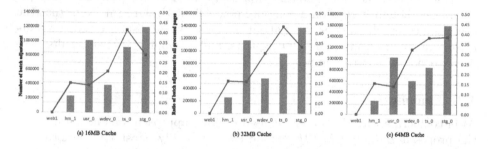

(a) 16MB Cache   (b) 32MB Cache   (c) 64MB Cache

Fig. 5. Statistics on batch adjustment with varied cache configurations.

Except for the trace of *web1*, *Cached-CFLRU* triggers certain batch adjustment operations after replaying other traces, corresponding to 13.4%–43.4% of all I/O requests. This fact discloses that *Cached-CFLRU* results in a large number of batch adjustment operations by also referring to spatial locality, which are the cause of the improvement on cache hits.

Overhead. The main space overhead of the proposed method is caused by three linked lists, which are responsible for doubly linking the node structures of cached pages, including two links, and logic page addresses. Because we have a limited number of data pages in the SSD cache, the space overhead of our approach is fixed. More importantly, the related work of *CFLRU* does require the same size of memory for holding the node information on the cached items, though it has only one doubly linked list.

Figure 6 presents the results of computation overhead of our proposed method and *CFLRU* with varied cache configurations. As shown in the figure, our method causes more time overhead by 28.3 ms on average, in contrast to the related work of *CFLRU*. This is because *Batched-CFLRU* brings about more operations on the linked lists due to batch adjustment in cache management. But we emphasize that, *Batched-CFLRU* results in time overhead less than 67.0 ms

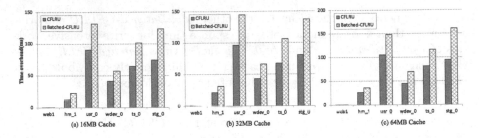

Fig. 6. Time overhead of batch adjustment in *Batched-CFLRU*.

after replaying the selected traces. This corresponds an average of $0.02\,\mu$s per I/O request, or less than 0.001% of the overall I/O time.

Thus, we conclude that the time overhead caused by our proposal is acceptable, even though it runs on a compute power-limited platform. Note that the computation overhead does bring about impacts on I/O response time by postponing dispatch on incoming I/O requests, and relevant I/O time results have been previously reported in Sect. 4.2

4.3 Summary

After comparing with the commonly used *CFLRU* approach, we understand that the proposed *Batched-CFLRU* method can noticeably boost I/O performance of SSD devices, with negligible space and time overhead. This is mainly because unifying both the factors of spatial locality and temporal locality in cache management of SSDs can minimize the negative effects of cache thrashing and thus enhance cache use efficiency of SSDs.

5 Conclusions

This paper presents a simple, effective cache management approach based on *CFLRU* for NAND-based SSDs, by uniting both temporal and spatial locality of references. To be specific, it holds three-level linked lists to shift the hot and cold buffered data pages. On an access of cached data page, it moves the corresponding node to the head of the highest-level linked list, and upgrade the neighboring nodes as the head(s) of the next highest-level linked list consequently. Besides, it removes the nodes from the tail end of the lowest level of linked list, to stand for the associated data pages are evicted from the cache. The experimental results illustrate our proposal can noticeably improve the measure of cache hits and then cut down I/O latency for SSDs, in contrast to the related work of *CFLRU*.

References

1. Liao, J., Zhang, F., Li, L., et al.: Adaptive wear-leveling in flash-based memory. IEEE CAL **14**(1), 1–4 (2015)

2. Kim, B., Choi, J., Min, S.: Design tradeoffs for SSD reliability. In: FAST (2019)
3. Liao, J., Gerofi, B., Lien, G., et al.: A flexible I/O arbitration framework for netCDF-based big data processing workflows on high-end supercomputers. In: CCPE 2017 (2017)
4. Cosmos OpenSSD Platform. http://www.openssd-project.org
5. Wang, Y.L., Kim, K.T., Lee, B., Youn, H.Y.: A novel buffer management scheme based on particle swarm optimization for SSD. J. Supercomput. **74**(1), 141–159 (2017). https://doi.org/10.1007/s11227-017-2119-2
6. Cui, J., Zhang, Y., Huang, J., et al.: ShadowGC: cooperative garbage collection with multi-level buffer for performance improvement in NAND flash-based SSDs. In: IEEE DATE (2018)
7. Li, J., Sha, Z., Cai, Z., et al.: Patch-based data management for dual-copy buffers in RAID-enabled SSDs. IEEE TCAD **39**, 3956–3967 (2020)
8. Chen, H., Pan, Y., Li, C., et al.: ECR: eviction-cost-aware cache management policy for page-level flash-based SSDs. CCPE **33**, e5395 (2019)
9. Sun, H., Dai, S., Huang, J., et al.: Co-active: a workload-aware collaborative cache management scheme for NVMe SSDs. IEEE TPDS **32**(6), 1437–1451 (2021)
10. Ari, I., Hong, B., Miller, E., et al.: Managing flash crowds on the internet. In: MASCOTS (2003)
11. Park, S.Y., Jung, D., Kang, J., et al. CFLRU: a replacement algorithm for flash memory. In: CASES (2006)
12. Khan, S.U., Yang, S., Wang, L., et al.: A modified particle swarm optimization algorithm for global optimizations of inverse problems. IEEE TOM **52**(3), 1–4 (2015)
13. Wang, M., Li, Z.: A spatial and temporal locality-aware adaptive cache design with network optimization for tiled many-core architectures. In: IEEE VLSI (2017)
14. Wang, H., Yi, X., Huang, P., et al.: Efficient SSD caching by avoiding unnecessary writes using machine learning. In: ICPP (2018)
15. Yoon, J., Won, W.: Access characteristic-based cache replacement policy in an SSD. In: ICCVW (2019)
16. Wu, S., Lin, Y., Mao, B., et al.: GCaR: garbage collection aware cache management with improved performance for flash-based SSDs. In: ICS (2016)
17. Hu, Y., Jiang, H., Feng, D., et al.: Exploring and exploiting the multilevel parallelism inside SSDs for improved performance and endurance. IEEE TC **62**(6), 1141–1155 (2013)
18. Ahmed, I., Sparsh, M., Mohammed, A., et al.: A survey of techniques for architecting SLC/MLC/TLC hybrid Flash memory-based SSDs. CCPE **30**(13), e4420 (2018)
19. Gao, C., Ye, M., Li, Q., et al.: Constructing large, durable and fast SSD system via reprogramming 3D TLC flash memory. In: MICRO (2019)
20. Search Engine I/O. http://traces.cs.umass.edu/index.php/Storage/Storage. Accessed Mar 2020
21. Narayanan, D., Donnelly, A., Rowstron, A.: Write off-loading: practical power management for enterprise storage. ACM TOS **4**(3), 1–23 (2008)

Networks and Communications

Taming Congestion and Latency in Low-Diameter High-Performance Datacenters

Renjie Zhou, Dezun Dong$^{(\boxtimes)}$, Shan Huang, Zejia Zhou, and Yang Bai

National University of Defense Technology, Changsha, China
{renjiezhou,dong,huangshan12,zhouzejia17,baiyang14}@nudt.edu.cn

Abstract. High-performance computing (HPC) and data centers are showing a trend of merging into high-performance data centers (HPDC). HPDC is committed to providing extremely low latency for HPC or data center workloads. In addition to adopting a low-diameter network topology, HPDC also requires a more advanced congestion control mechanism. This paper implements the state-of-the-art congestion control method in the data centers on the Dragonfly topology and proposes Bowshot, a fast and accurate congestion control method for low-latency HPDC. Bowshot uses fine-grained feedback to accurately describe the network state. It uses switch feedback, ACK-padding, and ACK-first to reduce feedback delay. Bowshot uses switch calculation to reduce the overhead of congestion control. As the large-scale evaluation shows, Bowshot reduced the average flow completion time (FCT) by 33% and the 99th percentile FCT by 45% compared to the state-of-the-art work. Bowshot reduces the feedback delay by 89%. In addition, Bowshot maintains higher throughput and a near-zero queue length.

Keywords: High-performance computing · Data center · Congestion control · Feedback accuracy · Feedback delay

1 Introduction

High-performance computing (HPC) and data centers are showing a trend of merging into high-performance data centers (HPDC). Their workloads are becoming more data-centric, and large amounts of data need to be exchanged with the outside world. HPDC uses an optimized Ethernet protocol at the link layer and uses RDMA technology (RoCEv2) [27] at the upper layer, which allows it to interoperate with standard Ethernet equipment while providing high performance for applications. Cray recently designed the SLINGSHOT [7] interconnection network which can provide good support for both HPC and data center.

The workloads of HPC and data centers are latency-sensitive and tail latency is crucial for such applications. They require the interconnection network to

© IFIP International Federation for Information Processing 2022
Published by Springer Nature Switzerland AG 2022
C. Cérin et al. (Eds.): NPC 2021, LNCS 13152, pp. 229–242, 2022.
https://doi.org/10.1007/978-3-030-93571-9_18

provide extremely low latency to ensure performance and user experience [5, 6, 8, 19]. The latency in the network can be divided into static latency and dynamic latency. Low-diameter network topology can provide very low static latency [13], e.g.the Dragonfly topology [15] has a diameter of 3 hops, which is cost-effective for HPDC networks. Also, efficient congestion control is of vital importance for HPDC to maintain low dynamic latency.

However, to meet the requirement of extremely low latency in HPDC networks, the existing congestion control mechanisms need to be improved. The standard mechanisms use ECN [10, 22] as the congestion signal. One bit ECN is a coarse-grained signal that cannot accurately reflect the network status. Due to the coarse-grained congestion signal, these algorithms require multiple iterations to converge to the optimal level. Therefore, these methods cannot accurately control network congestion in time, which is generally unacceptable for bursty HPDC workloads.

In recent years, the in-band network telemetry (INT) [14] is drawing much attention. INT is a framework designed to allow network data plane status to be collected and reported on the switch without any control plane intervention. It can be used for congestion control because it obtains fine-grained network status.

HPCC [18] is the state-of-the-art method in data centers community. It uses INT to obtain fine-grained network status to make a precise congestion control. We deployed HPCC in Dragonfly topology. Thanks to the fine-grained feedback, HPCC can control congestion well and reduce latency more effectively than traditional ECN-based methods. Meanwhile, we found that HPCC still suffers from untimely control in the deployment. First, there is a lack of control in the first RTT sent by the HPCC in the flow. Second, the sender cannot timely adjust the congestion due to the high feedback delay. In addition, HPCC uses the INT to pad the raw fine-grained information into the packet, which causes a lot of control overhead.

We propose a method for low-diameter HPDC called Bowshot, which can achieve low-overhead, timely, and precise congestion control. Bowshot uses fine-grained feedback for precise feedback. Mechanisms such as switch feedback, ACK-Padding, and ACK-first make feedback more timely. Bowshot uses switch calculations to reduce the transmission overhead of congestion control. Therefore, Bowshot can use lower control overhead to implement timely and accurate congestion control on low-diameter HPDC networks.

In general, this paper has the following four main contributions:

- We deployed the data center congestion control method HPCC in the Dragonfly topology and analyzed its limitations.
- This paper proposes a congestion control method for low-latency HPDC called Bowshot, which can control congestion in time and accurately.
- We have proposed methods such as switch feedback, ACK-padding, and ACK-first for Bowshot so that congestion can be controlled in a more timely manner.
- We develop switch calculation to reduce the overhead of congestion control.

We conduct a lot of experiments in Dragonfly topology, Bowshot reduces the average flow completion time (FCT) by 33% and the 99th percentile FCT by 45% compared to HPCC. Bowshot reduces the feedback delay by 89%. In addition, Bowshot maintains higher throughput and a near-zero queue length.

2 Background and Motivation

Both HPC and data center workloads require extremely low latency to ensure application performance and user experience. Low-diameter network topology can provide very low static latency (e.g. Dragonfly topology [15]). Existing ECN-based [10,22,23] congestion control methods need to be improved due to their inaccuracy and long feedback delay. Since the INT function can obtain fine-grained network status and has been used for high precise congestion control in the data center community (E.g. HPCC [18]), we deployed HPCC in the Dragonfly topology. However, we found that HPCC still suffers from untimely feedback.

2.1 Low-diameter Network Topology

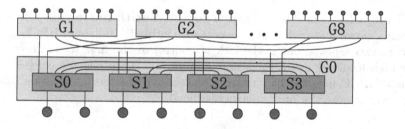

Fig. 1. An example Dragonfly topology.

One of the most popular interconnection network designs used in HPC and data centers is the fat-tree [17], which can provide very high binary bandwidth. However, due to the large network diameter of fat-tree, it not only brings higher network costs but also increases end-to-end delay. A variety of low-diameter network topologies have been proposed, which are more cost-effective than three-level fat trees and have lower latency. Among them, the most widely deployed is the dragonfly topology, such as the Cray Cascade system [9] and the Slingshot [7] system. Dragonfly is a hierarchical direct topology in which all switches are connected to compute nodes and other switches. The set of switches are connected to form a group, and groups are connected in a fully connected graph. Figure 1 shows an example, each group consists of 4 switches; there are a total of 9 groups in the topology.

2.2   ECN-Based Method and INT-Based Method

ECN is a mechanism in InfiniBand networks and is widely used in large network systems. When the network is congested, the data packet will be marked with an ECN mark when passing through the congested switch, the ACK will inherit the ECN mark of the data packet, and then the sender will reduce the sending rate according to the ECN mark in the ACK. However, ECN has only one bit and cannot accurately describe the network status. Therefore, they require multiple iterations to converge to the optimal level and cannot meet the low-latency requirements.

INT is a framework designed to allow the collection and reporting of the status on the switch(e.g. switch ID, link utilization, or queue status). The INT-based method uses this fine-grained information for precise congestion control. In HPCC, each data packet will be padded with INT information when passing through the switch. When the data packet arrives at the receiver, the INT information is copied to the corresponding ACK. Then ACK is sent back to the sender. Finally, the sender performs congestion control based on the INT information padded in the received ACK.

2.3   HPCC Suffers from Untimely Control and High Control Overhead

HPCC can achieve precise congestion control, but there are still problems of untimely control and high control overhead. Untimely control of HPCC is manifested in two aspects: one is the lack of control in the first RTT; Second, due to the high feedback delay, the sender cannot quickly adjust the congestion. In addition, HPCC uses the INT to pad the raw fine-grained information into the packet, which causes a lot of control overhead.

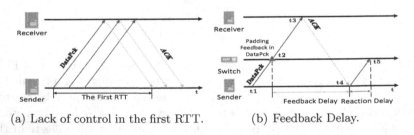

(a) Lack of control in the first RTT.   (b) Feedback Delay.

Fig. 2. Untimely control of HPCC is manifested in two aspects: (a) shows that HPCC lack feedback in the first RTT; (b) shows that the feedback delay of HPCC is too long.

HPCC Lack Control in the First RTT. Figure 2(a) shows that when the sender starts to send a flow, it needs at least one RTT to receive ACK. Therefore, each flow can send data at wire speed uncontrollably within the first RTT. As the link rate between HPC and data center gradually increases, uncontrolled data increases accordingly. This uncontrolled data can cause and aggravate network congestion. So we leverage the switch feedback to make up for the lack of control in the first RTT.

High Feedback Delay in the HPCC Flow Transmission. We consider the time required for feedback to take effect as the sum of feedback delay and response delay. The feedback delay is the time for the feedback transmission. The response delay is the time when the sender's control takes effect. Response delay is fixed, but feedback delay can be reduced. In HPCC, from $t2$ to $t5$, the control of the sender needs at least one RTT to take effect. Whether the link is idle or congested, it is difficult for HPCC to achieve fast convergence due to high feedback delay. Therefore, we use ACK-padding and ACK-first methods to reduce the feedback delay.

Raw Fine-Grained Information Causes a Lot of Overhead. HPCC uses the INT to pad the raw fine-grained information into the packet, which causes a lot of control overhead. In HPCC, each time a data packet passes through a switch, it is padded with 64-bit feedback data. In a network with a diameter of 3 hops, the feedback overhead padded in data packets and ACKs reaches 34 Bytes, which is a non-negligible overhead. We use the switch calculation method to process the raw fine-grained information on the switch before feedback, which significantly reduces the control overhead.

3   Design

In this section, we will introduce Bowshot, which includes the design principles and design details of Bowshot.

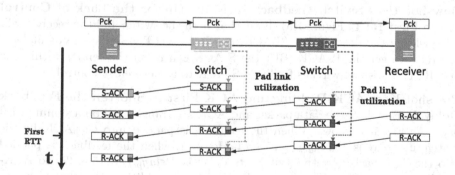

Fig. 3. The overview of Bowshot framework.

3.1 Design Overview

As shown in the Fig. 3, Bowshot is a sender-driven congestion control framework. It uses fine-grained feedback to accurately describe the network status. It uses switch feedback, ACK-padding, and ACK-first to reduce feedback delay. Bowshot uses switch calculation to reduce the overhead of congestion control, and the data packet and ACK are transmitted on a symmetrical path. S-ACK is used to make up for the lack of control in the first RTT. Bowshot switches use INT to obtain fine-grained network status and calculate link utilization as the feedback signal. It pads the feedback in the ACK instead of the data packet and gives ACK a higher priority to reducing the feedback delay. Senders multiplicatively adjust the sending window according to the maximum link utilization for fast convergence.

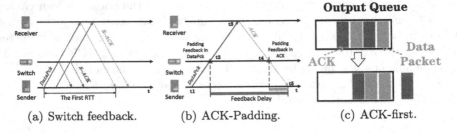

(a) Switch feedback.   (b) ACK-Padding.   (c) ACK-first.

Fig. 4. Design Rationale. (a) illustrates that switch feedback can feedback early in the first RTT; (b) and (C) illustrate that ACK-Padding and ACK-first can reduce the feedback delay.

Bowshot Uses Switch Feedback to Make Up for the Lack of Control in the First RTT. Figure 4(a) shows that both the switch and the receiver will acknowledge the data packet (S-ACK and R-ACK). The sender needs at least one RTT to get the R-ACK. BUt the S-ACK can reach the sender within the first RTT. Bowshot switches feedback S-ACK to achieve early control.

Bowshot Uses ACK-Padding and ACK-First to Shorten the Feedback Delay. In Bowshot, the data packet and ACK are transmitted on a symmetrical path. As Fig. 4(b) shows, When the feedback is padded into the ACK by the switch, its delay is $Delay_{ack} = t5 - t4$. However, when the feedback is padded into the data packet by the switch, its delay is $Delay_{pck} = t5 - t2$. So ACK-Padding can significantly reduce feedback delay. In addition, Bowshot also uses the ACK-first to further reduce the feedback delay. As shown in Fig. 4(c), when there are both ACK and data packets in the queue of the switch, ACK is sent first to ensure timely feedback.

Bowshot Uses Switch Calculation to Reduce Control Overhead. The Bowshot switch calculates the fine-grained status information into link utilization, and then sends the link utilization as feedback to the sender. The link utilization can be represented by a 16-bit half-precision floating-point number. Bowshot pads the maximum link utilization in the path into the ACK, so the overhead of each ACK of Bowshot is fixed length and does not change with the topology. In addition, in HPCC, both data packets and ACKs need to be padded with feedback. However, Bowshot uses the ACK-Padding method, so only ACK needs to be padded with feedback, which further reduces overhead.

3.2 Design Details

In this section, we describe the design details of Bowshot, including switch design and endpoint design.

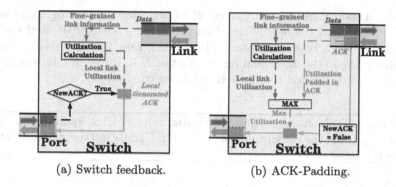

(a) Switch feedback.   (b) ACK-Padding.

Fig. 5. (a) illustrates that the Bowshot switch conditionally generates an ACK for the received packet. (b) illustrates that the switch pads the maximum link utilization into the received ACK.

Switch Design

Link Utilization Calculation: Bowshot uses link utilization U as a feedback signal in the network. Bowshot switch uses the fine-grained data link information from INT to calculate the U, then, the switch pads U into the ACK it generates or receives. The U is calculated as follows:

$$U = \frac{qlen + RTTbytes}{RTT * B} \tag{1}$$

where $qlen$ is the queue length of the output port, and $RTTBytes$ is the amount of data sent by the port in the last RTT, so $qlen + RTTbytes$ is the total data that the link needs to transmit in one RTT. B is the link's bandwidth, and $RTT * B$ is the maximum data that the link can send in one RTT. Therefore, the U is the ratio of the data volume that needs to be transmitted to the transmission capacity in one RTT, representing the congestion level of the link.

Switch Feedback: In the Bowshot switch, whether each active stream generates an S-ACK depends on the boolean value $NewACK$, and its initial value is *true*.

As shown in Fig. 5(a), when the switch receives a data packet and $NewACK$ is $true$, the switch will generate the corresponding S-ACK, which contains the local link utilization. As shown in the Fig. 5(b), when the switch receives ACK, S-ACK is no longer generated, and $NewACK$ is set to $false$. Therefore, each switch only needs to generate a few S-ACKs for feedback in the first RTT.

ACK-Padding: Bowshot switch pads the maximum link utilization into the received ACK. As Fig. 5(b) shows when the switch receives the ACK, it calculates the local link utilization U_{local}, then the switch compares U_{local} with U_{ack} padded in ACK and pads the largest one into ACK. Therefore, when the sender receives an ACK, the link utilization padded in the ACK represents the most congested link that the ACK has passed.

Endpoint Design. Algorithm 1 shows the congestion control logic of a single flow. The sending of the flow is based on the sending window W, and the flow stops sending when W is exhausted. The sending rate R is adjusted according to W. According to the maximum link utilization U in the ACK, the sender uses the multiplicative method to adjust the sending window.

Algorithm 1. Endpoint algorithm. The initial value of W is equal to bandwidth multiplied by RTT, the initial value of the $SACKUpdate$ is True, the initial value of W^c is the same as W.

```
1: function HANDLEACK(ack)
2:     U = ack.U
3:     W = Wᶜ/(U/η) + W_AI
4:     if ack is S-ACK and SACKUpdate is True then
5:         Wᶜ = W
6:         SACKUpdate = False
7:     else
8:         if ack.seq > lastUpdateSeq then
9:             Wᶜ = W
10:            lastUpdateSeq = snd_nxt
11:    R = W/T
```

η is the ideal link utilization. So the relationship between the ideal sending window and the current window is shown in Eq. (2).

$$\frac{W_{ide}}{W_{cur}} = \frac{\eta}{U} \tag{2}$$

As shown in Eq. (3). Bowshot sender uses MIMD to quickly adjust the sending window W. In order to avoid being unable to restart after the sending window drops to zero, the algorithm needs to add a small increment W_{AI}. In order to ensure that the link capacity will not be oversubscribed, the small increment is usually set to $BDP * (1 - \eta)/n$, N is the maximum number of flows that pass the link at the same time.

$$W = \frac{W^c}{U/\eta} + W_{AI} \tag{3}$$

Continuous ACKs will carry overlapping network feedback, and the sender cannot perform multiplicative operations on all ACKs, which will lead to over-reaction. As shown in Algorithm 1, every time the sender receives an ACK, it calculates the sending window W according to the reference window W^c. The variable $lastUpdateSeq$ records the sequence number of the first packet sent after W^c is updated. The sender updates W^c after receiving the ACK corresponding to the record number. In the first RTT, We use the variable $SACKupdate$ to ensure that the W^c updates only once.

4 Evaluation

In this section, we conducted a lot of experiments to evaluate Bowshot's performance. These experiments are performed on the OMNeT++ simulator [2] and our extended INET framework [1].

4.1 Evaluation Configuration

Network Topologies. Most of the evaluations in this paper used the Dragonfly topology with 342 endpoints. There are 6 switches in each group, each switch is connected to 3 endpoints, and each switch has 3 global links connected to switches in other groups. In the topology, we use shortest path routing. The rate of all links in the topology is 100 Gbps, and the link delay between switches is 100 ns. Due to the PCIe delay, the delay between the switch and the endpoint is 600 ns.

Traffic Patterns. We carefully evaluated the data center and HPC scenarios. For the HPC scenario, we evaluated the CG benchmark in the NAS Parallel Benchmark (NPB) [4]. For the data center, we evaluated two scenarios: Cache Follower [21] and Web Search [3]. They are widely recognized and open test standards. We set the average link load to 0.4, 0.6, and 0.8, respectively. In addition, we also used some simple manual scenarios to evaluate Bowshot's microbenchmarks.

Performance Metrics. In the evaluation, we mainly compare the following four metrics: (1) the delay of network feedback; (2) the completion time of the flow or message (FCT); (3) the queue length of the switch. (4) The throughput of the endpoint;

4.2 Micro-benchmarks

In this section, we evaluate the fairness of Bowshot and compare the feedback delays of Bowshot and HPCC.

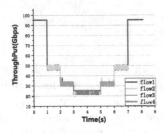

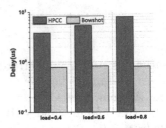

Fig. 6. Fair-share of Bowshot.

Fig. 7. Average feedback delay in the path.

Fairness: In a many-to-one topology, flow1 to flow4 sends data to the receiver one by one at an interval of 1 s, and then stops sending one by one. As shown in Fig. 6, Bowshot can achieve good fairness. When new flows join the link, all flows can quickly converge to a fair bandwidth. When there is idle link capacity, each flow can quickly and fairly occupy the idle link capacity.

Feedback Delay: We ran three data center or HPC scenarios in the 342-nodes Dragonfly topology on the 0.8 workloads. We measure the feedback delay of each switch in the longest path and average them to get the feedback delay of the path. As Fig. 7 shows, The path feedback delay of HPCC is all above 4 μs and increases with the increase of load. However, Bowshot's feedback delay is only 800 ns, and it will not change with traffic patterns. Thanks to the ACK-padding and ACK-first methods, Bowshot reduced the feedback delay by 89%.

4.3   Large-Scale Simulations

In this section, we run three HPC or data center application scenarios with different workloads in a dragonfly topology with 342 endpoints to evaluate Bowshot's performance.

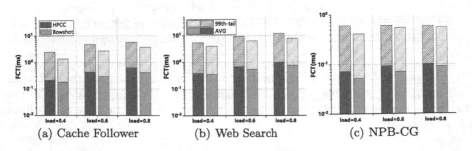

(a) Cache Follower   (b) Web Search   (c) NPB-CG

Fig. 8. Three methods of AVG FCT and 99-th FCT in different scenarios and workloads.

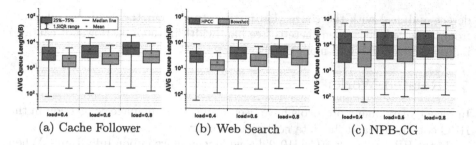

Fig. 9. Average queue distribution of all ports of switches in Dragonfly topology with different scenarios and workload.

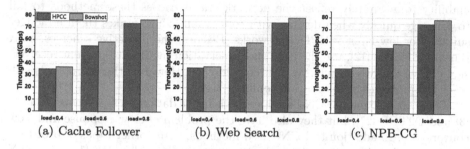

Fig. 10. Average throughput of all endpoint ports in Dragonfly topology with different scenarios and workload.

Bowshot can Significantly Reduce the FCT. We evaluated the 99th percentile FCT and average FCT of Bowshot and HPCC in data center and HPC scenarios. As Fig. 8 shows, Bowshot is better than HPCC in all scenarios. Tail latency is critical for data centers or HPC workloads. Bowshot can significantly reduce the tail latency. In the Cache Follower traffic pattern with a load of 0.4, the 99th-FCT of HPCC and Bowshot is 2.46 ms and 1.35 ms. Bowshot reduces the 99th-FCT by 45%, it can also reduce AVG-FCT by 33%. Bowshot can also reduce the tail delay by up to 31% and the average FCT by 26% in the HPC workload NPB-CG.

Bowshot can Achieve Near-Zero Queuing. We evaluated the average queue distribution of all ports of the switch in the Dragonfly topology. As shown in the Fig. 9, compared with HPCC, Bowshot can maintain a lower port queue length in different scenarios and workloads. In Bowshot, most of the port queues in all scenarios are below 10 KB equivalent to the size of a few packets. When the link is congested, the Bowshot sender can detect and control the congestion earlier. However, HPCC cannot detect congestion in time, which can easily cause congestion to worsen.

Bowshot can Maintain Higher Throughput. Figure 10 shows the comparison between HPCC and Bowshot on the average throughput of all endpoints in the topology. Regardless of HPC or data center workloads, Bowshot can increase throughput by 4% to 6%. This is because when congestion occurs, Bowshot can detect and deal with the congestion in time to prevent the sending end's rate from being reduced; when there is free bandwidth in the link, Bowshot can also increase its sending rate more timely.

5   Related Works

In recent years, many protocols have been proposed to manage congestion in the HPC community and the data center community.

In the HPC network, ECN [10,22] is widely employed upon InfiniBand. When the network is congested, switches will mark ECN on the packet, and the sender will reduce the injection rate when receives the marked feedback. However, the inability to accurately reflect the network status causes these methods to fail to converge quickly when dealing with congestion, which makes them no longer suitable for low-latency HPDC networks. Besides, the proactive schemes, SPR [11] and SMSRP [12], use the time slice reservation method to prevent receiver-side congestion, which achieves fast convergence. However, these protocols still meet the problems of reservation conflicts [24].

There are also many ECN-based methods in the data centers, such as DCTCP [3] and D2TCP [27], but they also have the problem of slow response and slow convergence. In addition to ECN-based methods, many other types of protocols have emerged in data center networks. Some protocols, such as DX [16], TIMELY [20] control congestion based on the delay, they adjust the sender's injection rate according to the queuing delay or end-to-end delay in the network. However, these delay-based methods usually do not perform well due to inaccurate delay measurement.

The INT [14] function can obtain fine-grained status information on the switch data plane, which can be used for precise congestion control. HPCC [18] uses the raw fine-grained state provided by INT for precise congestion control, but HPCC still has the problems of untimely control and high control overhead. APCC [26] and FastTune [25] two methods alleviate the problem of untimely HPCC control to a certain extent. Neither APCC nor FastTune solves the problem of high HPCC control overhead, and APCC even increases control overhead. Bowshot uses switch feedback, ACK-Padding, ACK-first and Switch calculation methods to provide low-latency congestion control for HPDC networks.

6   Conclusion

This paper implements the state-of-the-art data center method on the Dragonfly topology and proposes Bowshot, a fast and accurate congestion control method for low-latency HPDC. It leverages fine-grained feedback to make precise congestion control. Bowshot uses switch feedback, ACK-padding, and ACK-first to realize timely control. It also uses the switch calculation to reduce the overhead of fine-grained feedback. Compared with the state-of-the-art work, Bowshot reduces the average flow completion time (FCT) by 33% and the 99th percentile FCT by 45%. Bowshot reduces the feedback delay by 89%. In addition, Bowshot maintains higher throughput and a near-zero queue length.

Acknowledgements. We would like to thank the anonymous reviewers for their insightful comments. The work was supported by the National Key R&D Program of China under Grant No. 2018YFB0204300, Excellent Youth Foundation of Hunan Province (Dezun Dong) and National Postdoctoral Program for Innovative Talents Grant No. BX20190091.

References

1. INET (2021). https://inet.omnetpp.org/
2. OMNeT++ (2021). https://omnetpp.org
3. Alizadeh, M., et al.: Data center TCP (DCTCP). In: Proceedings of the ACM SIGCOMM (2010)
4. Bailey, D., Harris, T., Saphir, W., Van Der Wijngaart, R., Woo, A., Yarrow, M.: The NAS parallel benchmarks 2.0. Technical report, Technical Report NAS-95-020, NASA Ames Research Center (1995)
5. Chunduri, S., et al.: GPCNeT: designing a benchmark suite for inducing and measuring contention in HPC networks. In: Proceedings of the SC (2019)
6. De Sensi, D., Di Girolamo, S., Hoefler, T.: Mitigating network noise on dragonfly networks through application-aware routing. In: Proceedings of the SC (2019)
7. De Sensi, D., Di Girolamo, S., McMahon, K., Roweth, D., Hoefler, T.: An in-depth analysis of the slingshot interconnect (September 2020)
8. Dean, J., Barroso, L.A.: The tail at scale. Commun. ACM **56**(2), 74–80 (2013)
9. Faanes, G., et al.: Cray cascade: a scalable HPC system based on a dragonfly network. In: Proceedings of the SC (2012)
10. Floyd, S.: TCP and explicit congestion notification. ACM SIGCOMM Comput. Commun. Rev. **24**(5), 8–23 (1994)
11. Jiang, N., Becker, D.U., Michelogiannakis, G., Dally, W.J.: Network congestion avoidance through speculative reservation. In: Proceedings of the ISCA (2012)
12. Jiang, N., Dennison, L., Dally, W.J.: Network endpoint congestion control for fine-grained communication. In: Proceedings of the SC (2015)
13. Kathareios, G., Minkenberg, C., Prisacari, B., Rodriguez, G., Hoefler, T.: Cost-effective diameter-two topologies: analysis and evaluation. In: Proceedings of the SC (2015)
14. Kim, C., Sivaraman, A., Katta, N., Bas, A., Dixit, A., Wobker, L.J.: In-band network telemetry via programmable dataplanes. In: Proceedings of the ACM SIGCOMM (2015)
15. Kim, J., Dally, W.J., Scott, S., Abts, D.: Technology-driven, highly-scalable dragonfly topology. In: Proceedings of the ISCA (2008)
16. Lee, C., Park, C., Jang, K., Moon, S., Han, D.: Accurate latency-based congestion feedback for datacenters. In: Proceedings of the USENIX ATC (2015)
17. Leiserson, C.E.: Fat-trees: universal networks for hardware-efficient supercomputing. IEEE Trans. Comput. **100**(10), 892–901 (1985)
18. Li, Y., et al.: HPCC: high precision congestion control. In: Proceedings of the ACM SIGCOMM (2019)
19. Misra, P.A., Borge, M.F., Goiri, Í., Lebeck, A.R., Zwaenepoel, W., Bianchini, R.: Managing tail latency in datacenter-scale file systems under production constraints. In: Proceedings of the EuroSys (2019)
20. Mittal, R., et al.: Timely: RTT-based congestion control for the datacenter. In: Proceedings of the ACM SIGCOMM (2015)

21. Roy, A., Zeng, H., Bagga, J., Porter, G., Snoeren, A.C.: Inside the social network's (datacenter) network. In: Proceedings of the ACM SIGCOMM (2015)
22. Santos, J.R., Turner, Y., Janakiraman, G.: End-to-end congestion control for infiniband. In: Proceedings of the IEEE INFOCOM (2003)
23. Thaler, P.: IEEE 802.1Qau congestion notification (2006)
24. Wu, K., Dong, D., Li, C., Huang, S., Dai, Y.: Network congestion avoidance through packet-chaining reservation. In: Proceedings of the ICPP (2019)
25. Zhou, R., Dong, D., Huang, S., Bai, Y.: FastTune: timely and precise congestion control in data center network. In: Proceedings of the IEEE ISPA (2021)
26. Zhou, R., Yuan, G., Dong, D., Huang, S.: APCC: agile and precise congestion control in datacenters. In: Proceedings of the IEEE ISPA (2020)
27. Zhu, Y., et al.: Congestion control for large-scale RDMA deployments. In: Proceedings of the ACM SIGCOMM (2015)

Evaluation of Topology-Aware All-Reduce Algorithm for Dragonfly Networks

Junchao Ma, Dezun Dong$^{(\boxtimes)}$, Cunlu Li, Ke Wu, and Liquan Xiao

National University of Defense Technology, Changsha, China
{majunchao,dong,cunluli,wuke13,xiaoliquan}@nudt.edu.cn

Abstract. Dragonfly is a popular topology for current and future high-speed interconnection networks. The concept of gathering topology information to accelerate collective operations is a very hot research field. All-reduce operations are often used in the research fields of distributed machine learning (DML) and high-performance computing (HPC), because All-reduce is the key collective communication algorithm. The hierarchical characteristics of the dragonfly topology can be used to take advantage of the low communication delay of adjacent nodes to reduce the completion time of All-reduce operations. In this paper, we propose g-PAARD, a general proximity-aware All-reduce communication on the Dragonfly network. We study the impact of different routing mechanisms on the All-reduce algorithm, and their sensitivity to topology size and message size. Our results show that the proposed topology-aware algorithm can significantly reduce the communication delay, while having little impact on the network topology.

Keywords: All-reduce operation · Dragonfly topology · Collective communication

1 Introduction

Dragonfly are typically deployed in many high-performance computer systems, including the Cray Cascade system [7], Titan [6] and Trinity [3]. The Cray Slingshot network is designed for continuous computation and also uses dragonfly topology [14].

Recently, it is popular to continuously improve the performance of collective operations so that it can run efficiently on various hardware and software platforms. All-reduce aggregates the values of all processes and then sends the values back to all nodes. All-reduce simplifies a complex set of point-to-point communications, making it easier for programmers to perform parallel and distributed programming. In addition, All-reduce operations can effectively separate application and interface developers, and contribute to the portability of functions and performance between applications and interfaces.

© IFIP International Federation for Information Processing 2022
Published by Springer Nature Switzerland AG 2022
C. Cérin et al. (Eds.): NPC 2021, LNCS 13152, pp. 243–255, 2022.
https://doi.org/10.1007/978-3-030-93571-9_19

There are already many All-reduce implementation algorithms in the mpi library. According to the number of processes and the size of the message, different algorithm choices can speed up the efficiency of communication. The most popular All-reduce algorithms are the ring algorithm and the halving and doubling algorithm. Ring algorithm [15] is an algorithm that makes full use of bandwidth. However, it is mainly suitable for tasks consisting of long messages because the number of steps executed increases linearly as the number of processes increases. Conversely, the halving and doubling algorithm [16] performs the least communication steps and achieves the least delay. While delay-sensitive and tasks consisting of small messages often use the halving and doubling algorithm, it suffers from considerable bandwidth overhead. There are also several hybrid methods, which complete the operations by combining different types of sub-step operations, including reduce-scatter and All-gather, as well as reducing and broadcasting [13].

G-PAARD is designed as a topology-aware algorithm to accelerate the All-reduce communication. The communication mode of the existing acceleration algorithm implemented in the dragonfly network may cause network congestion and waste of link resources[1]. The key idea behind g-PAARD is to decompose All-reduce operations into reduce-scatter and All-gather modes in a topology-aware way. The All-reduce operation with g-PARRD can be completed in just 6 steps. Therefore, g-PAARD utilizes local communication with neighbors to minimize global communication and reduce overhead. Experimental results show that g-PAARD is superior to these state-of-the-art solution in a particular context.

The main contributions of this design can be summarized as follows:

- We have shown that existing algorithms are not well suited for Dragonfly networks, resulting in poor performance and limited overall network throughput.

- We propose g-PAARD, which permits the scheduling of an end-to-end solution to alleviate congestion.

- A comprehensive evaluation of the proposed algorithm has been performed and demonstrated that higher performance can be achieved in a particular context.

2   Motivation

2.1   Dragonfly

Kim et al. [10] introduced Dragonfly networks, it has become one of the most popular topologies in high cardinality HPC interconnect networks. Dragonfly has the advantages of high scalability, small diameter and low cost. The standard Dragonfly topology employs a two-tier structure, as shown in the Fig. 1.

---

[1] For example, using standard algorithms, both nodes require at least 3 hops, and in some cases up to 6 hops, to facilitate communication at each step of the topology.

The routers are interconnected using a fully connected intra-group topology to create a group. A palmtree for $a = 4$ is included in Fig. 1. The number of links connecting each router to the local compute node is p; the number of routers in each group is a; the number of global links each router connects to routers in other groups is h; and the number of groups be g.

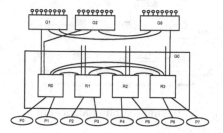

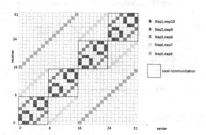

Fig. 1. An example Dragonfly topology with 9 groups and each group contains 4 routers. DF($p = 2$, $a = 4$, $h = 2$, $g = 9$)

Fig. 2. Communication pattern of HD for 32 nodes in DF(2,4,2) network.

In the dragonfly topology, data packets are routed along the minimum or non-minimum path. The minimum path is any path from the source node to the target node, and contains at most one global link. The smallest path takes one hop in the source group (from the source switch to the switch with the global link to the destination group), then travel through the global link to the destination group, and finally selects the local link to the destination at the target group. Depending on the source location and target location, the minimum path may have fewer hops.

The Minimal routing (MIN) scheme routes packets only through minimal paths, thus minimizing resource usage. MIN works well for traffic patterns where the load can be evenly distributed, such as random uniform traffic. However, as the number of links between each pair of groups is small, MIN routing performs poorly for adversarial traffic, particularly where most communication arises between two groups.

Generally, the delay of the global link is greater than the delay of the local link. In the dragonfly topology, there are some pairs of nodes connected by global links between different groups. These node pairs can communicate at the same time to maximize the utilization efficiency of the global link. In this article, we will consider the characteristics of Dragonfly to design an efficient All-reduce algorithm on the Dragonfly network. This can significantly reduce communication time by coordinating communication between adjacent nodes and completing the All-reduce operation.

2.2 Existing All-Reduce Algorithms

– Ring Algorithm (Ring)

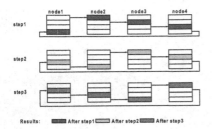

Fig. 3. Ring algorithm for all-reduce

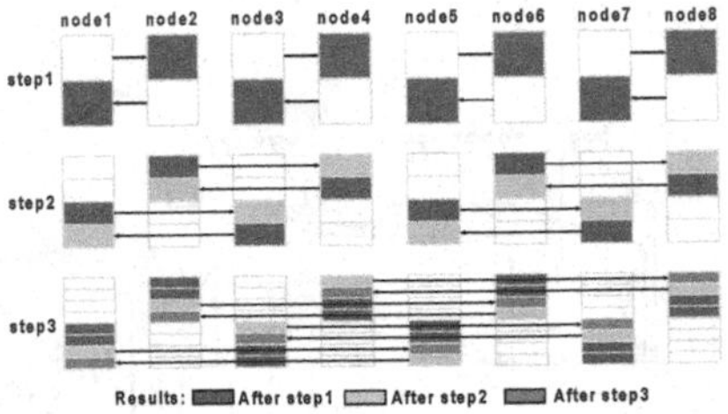

Fig. 4. Halving-doubling algorithm for all-reduce

Figure 3 present a ring algorithm for All-reduce. Utilizing chunked data transfer and reduction can optimize the ring algorithm. The entire ring process is divided into two major steps: the first step is scatter-reduce, and the second step is All-gather. First we have n nodes, then we divide the data (equal) on each node into n blocks, and assign each node its left and right neighbors (the left neighbor of node 1 is node 0, and the right neighbor is node 2...), and then start to perform $n-1$ operations. At the i-th operation, the node j will send its $(j-i)\%n$-th block of data to node $j+1$, accept the $(j-i-1)\%n$ block of data from node $j-1$, and perform a reduce operation on the received data. When $n-1$ operations are completed, the first scatter-reduce step of ring-allreduce has been completed. At this time, the $(i+1)$-th node of the i-th node $\%$ n blocks of data have collected the $(i+1)\%n$ block of all n nodes.

Then, perform All-gather again to complete the algorithm. The second phase for All-gather is very simple. It is to pass the $(i+1)\%$ n-th block of the i-th node to other nodes through $n-1$ passes, and it is also in the i-th pass. Node j sends its $(j-i-1)\%n$-th block of data to the right neighbor, and accepts the $(j-i-2)\%n$-th data of the left neighbor, but the received data does not need to be like the first step to reduce, but directly replace its own data with the received data.

The ring algorithm [15] can make full use of bandwidth and is suitable for large message tasks. However, as the number of nodes increases, the number of steps executed by the algorithm increases linearly. This greatly increases the delay time of small message communication. In the dragonfly topology, each step of the ring algorithm needs to use the global link for communication, which will further increase the complete time.

– Halving-doubling Algorithm

Figure 4 gives an example to show the All-reduce process of the HD algorithm. The entire HD process is divided into two major steps, the first stage is scatter-reduce, and the second stage is All-gather. In the first step, node 0 and the node with a distance of 1 (node 1) exchange half of the data aggregated in the previous step (the first step) and aggregate. In step $\log_2 P$, node 0 and the node at distance 2^{n-1} exchange half of the data aggregated in the previous step $\log_2(P-1)$ and aggregate. The second stage is the reverse process of the first stage. In the first step, node 0 exchanges aggregated data with a node at a distance of 2^{n-1}. In the second step, node 0 exchanges updated aggregated data with the node at a distance of 2^{n-2}. The node 0 and the node with a distance of 1 (node 1) exchange half of the aggregated data. Finally, all nodes get aggregated data, and the Allreduce operation is completed.

The HD algorithm is suitable for small message tasks and the number of nodes is a power of 2. At this time, the HD method greatly reduces the communication delay because of the few communication steps ($2\log_2 P$). However, in the dragonfly topology, since the distance between nodes may be 3 hops, each step of communication will have additional global link delay overhead. This will bring performance degradation. At the same time, the size of dragonfly topology is usually not a power of 2, which will add an extra step of communication overhead.

As shown in Fig. 2, it is a communication pattern of HD for 32 nodes in DF(2,4,2) network. Horizontal axis represents the sending node, and the vertical axis represents the receiving node. Different colors represent different steps. Communications carried out at the same time has been marked with the same color. It can be seen that nearly half of the communication is adversarial traffic, where nodes in one group should communicate with nodes in another group. These communications take multiple hops and add network contention.

3   The g-PAARD Design

3.1   g-PAARD Algorithm in Dragonfly

Notation: In the following we use the term *global node* to refer to the node connecting with global link directly. We call the *local node* to refer to the node connecting with global link indirectly.

Node Placement: The algorithms presented hereafter are based on the assumption that some information about the topology and the node placement can be obtained. Each node, router, and group is assigned a unique node, router, and group *id*, respectively. Any process can have access to the node *id* of any other process belonging to the same application.

g-PAARD Algorithm consist of six steps. First, the data of the node is evenly divided into g(which is the size of group) parts. *Local node* send specific $1/g$ data to *global node*. *Global node* send specific $1/g$ data to other *global node* and, *Global node* receive and reduce data from *local node* and other *global node* simultaneously. In the second step, *global node* in different group communication

248   J. Ma et al.

with each other through the directly connected global link. *Global node* send specific data to the pair *global node*. Simultaneously, every *global node* receive and reduce the specific data. In the third step, the aggregation data in second step is evenly divided into $p \times a$ (which is the size of nodes in a group) parts. Nodes in the same group communicate with each other. Node send specific data to other nodes in the same group. Simultaneously, every node receive and reduce specific data from other nodes in the same group. After third step, every node aggregate the $\frac{1}{g \times p \times a}$ (which is the number of all nodes). The reduce-scatter phase is completed. The next three step is reversed to complete the All-gather phase.

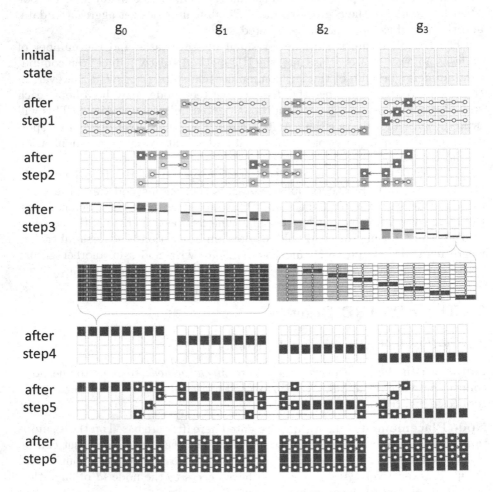

Fig. 5. g-PAARD for a 32 nodes in DF(2,4,2) networks

3.2 An Example of the g-PAARD Algorithm for 32 Nodes in DF(2,4,2) Network

As shown in Fig. 5, it is an example of the g-PAARD algorithm for 32 nodes in DF(2,4,2) network. We assume that 32 nodes are allocated in four groups, there are 8 nodes in a group. And a global palmtree arrangement is deployed [4].

g-PAARD requires two processing phases, each with several communication steps. As shown in 5, 8 columns per group represent data in 8 nodes, and 32 columns are given to represent 32 nodes. Due to the existence of four groups, the data in each node is divided into four parts (marked as a first to fourth data set). At first, each node n has its own local data.

In Fig. 5, we take nodes in $g0$ as an example to show the All-reduce process of g-PAARD. In the first step, for sender, *local node*(data are marked as blank square) sends the data set to *global nodes*(some data are marked as colored square) in $g0$, and *global nodes* sends the data set to other *global nodes* in $g0$. For the receiving end, the *global nodes* receive and aggregate data sets from all other nodes simultaneously. Remarkably, this kind of communication between nodes is direct and requires only one hop. After step 1, one quarter of the data in each node has not been transferred because it will be aggregated in step 2.

In the second step, communication takes place between the global groups. There are 6 global links between groups, so that 6 pairs of nodes can communicate at the same time. These communications can be accomplished with only a hop at the same time. Take the *global nodes* in $g0$, which sends aggregated data after step 1 to their counterparts (data marked green in other groups). At the same time, *global nodes* receive and aggregate data sets from their pairs. After step 2, all four groups received their specific data. Therefore, another step needs to be taken to complete the reduce-scatter phase.

Because there are eight nodes in a group, the data aggregated after step 2 can be divided into eight groups in step 3. Taking the node in $g0$ as an example again, the first set of data for each node in $g0$ is divided into eight parts (1, i (1~8)), where node 1 sends (1, i) data to node i in $g0$. At the same time, node 1 receives and aggregates data from other nodes in $g0$. After the third stage, the reduce-scatter phase of the data was completed, with 32 nodes collecting the data respectively. The All-gather phase is similar to the process leading to the final result. As a result, the g-PAARD algorithm reduces the dependent length and hops per step compared to the Ring algorithm and the HD algorithm, as shown in Table 1.

The next three step is a reversed phase to complete All-reduce operation. In the fourth step, Taking nodes in $g0$ as an example again, node 1 sends the aggregates data after step 3(marked as red) to other nodes in $g0$. Simultaneously, node 1 receives the aggregated data from other nodes in $g0$. After step 4, every node has 1/4 aggregated data.

In fifth step, *global nodes* in $g0$ send the aggregated data(marked as red) to their pair in other group, simultaneously *global nodes* receive aggregated data from *global nodes* in other group.

In sixth step, it is similar with step 1. Taking nodes in $g0$ as an example, *global nodes* send the data received in step 5 to other nodes in $g0$. After step 6, all nodes get the completed aggregated data. The All-reduce operation finish. While our example uses 32 nodes, the g-PAARD algorithm is applicable for arbitrary nodes counts.

As shown in Fig. 6, it is communication pattern of g-PAARD for 32 nodes in DF(2,4,2), horizontal axis represents the sending node, and the vertical axis represents the receiving node. Different colors represent different steps. Communications carried out at the same time has been marked with the same color. All the communication are proximity, most of communications occur in intra-group. thus, it reduce the global link latency compare to HD algorithm. And, The communications marked with blue occur in inter-group, and only take one hop. So, it can alleviate the global link contention.

Table 1. g-PAARD achieves good tradeoffs by minimizing hops and length of dependency

Algorithm	Ring	Halving doubling	g-PAARD
Minimum hop	1	3	1
No. of dependant steps	2(P−1)	2logP	6

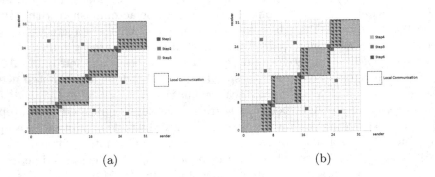

(a)   (b)

Fig. 6. Communication pattern of g-PAARD for 32 nodes in DF(2,4,2) network

4   Evaluation

4.1   Evaluation Methods

Our simulator is based on the Booksim router model, which emphasizes the precise simulation of the cycle of hardware components. In our evaluation, the accuracy of network simulations was at the microchip level. We expanded Booksim to support parallel distributed simulation for large scale networks to overcome the drawbacks of Booksim serial execution. In addition, we extend it to support parallel distributed emulation and model a complete MPI library (such as an SST emulator) to evaluate the full emulation of the entire process from MPI Call to generate network traffic.

4.2   Evaluation Results

A. Node Allocation

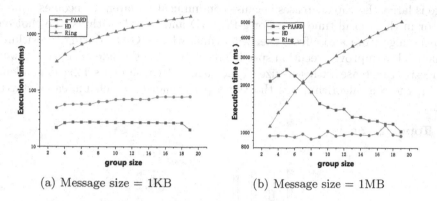

(a) Message size = 1KB

(b) Message size = 1MB

Fig. 7. Execution time when All-reduce 1 KB and 1 MB on a (3,6,3) 342 nodes Dragonfly network with minhop routing strategies. The job runs alone on the network

Figure 7 show the execution time when All-reduce 1 KB and 1 MB on a (3,6,3)342 nodes Dragonfly network with minhop routing strategies. When message size is 1 KB, g-PAARD perform better than HD and Ring algorithm. However, when message size is 1 MB, g-PAARD perform worse in small group size. g-PAARD relies more on a complete network to play its performance advantages. When group size is small, the global link decrease. g-PAARD lost the advantage of using all global links at the same time.

B. Router Algorithm

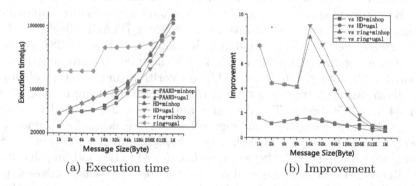

(a) Execution time

(b) Improvement

Fig. 8. Performance of three algorithms for 32 nodes in DF(2,4,2) 72 nodes

The performance of three algorithms is evaluated using minhop routing algorithm and UGAL routing algorithm. The topological scale is 32 nodes of Dragonfly(2,4,2) 72 nodes, and the result is shown in Fig. 8. When the message size

is between 1 k and 256 k, the performance of the g-PAARD algorithm is significantly better than the HD algorithm or the ring algorithm, because link latency
has a greater weight in the total time cost. However, when the size of a message is large, the gap decreases because communication latency becomes a major
factor in the overall time cost. For g-PAARD and HD algorithms, the choice of
routing algorithm also affects execution time, while UGAL routing algorithm is
beneficial to improve execution speed. This is because the side effects of network
congestion increase with the size of messages. Because of its long dependence
chain, the ring algorithm has the worst performance, which increases startup
delay.

C. Topology Scale

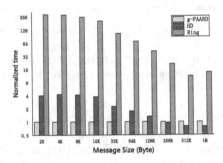

Fig. 9. Normalized time of three algorithm with minhop route algorithm in DF(4,8,4)
1056 nodes

Figure 9 illustrates the normalized time of the three algorithms using the minhop
route algorithm in DF(4,8,4) 1056 nodes. The data size was varied only from 2 K
to 1 MB in order to ensure the ring algorithm could be run. Figure 9 presents the
normalized time of the three algorithms. Here, the g-PAARD displays a 4.33×
speed increase over the halving-doubling algorithm at 4 kB and a 287.34× speed
increase over the ring algorithm at 2KB with the minhop route algorithm. As
compared with both the halving-doubling algorithm and the ring algorithm,
Fig. 9 illustrates the normalized time of three algorithms using minhop routing
in DF (4,8,4) 1056 nodes. The data size ranges from 2 K to 1 MB to ensure
that the ring algorithm can run. Figure 9 shows the normalized time of the
three algorithms. Here, g-PAARD shows accelerated growth 4.33× over recursive
doubling algorithm and 287.34× over ring. Compared to HD and ring algorithms,
g-PAARD reduces the time required more as the topology scale increases, as the
cost of link latency has a higher weight when node count is larger. Nevertheless,
when data sizes are particularly large, the performance of the three algorithms
becomes generally similar, as the communication cost has a higher weigh. When
the topology scale is larger, the lengthy of the dependency chain has greater
weight in terms of the cost.

5 Related Works

The approach of gathering topology information and leveraging this knowledge to design better algorithms has been applied to many systems in the past. For example, HAN [11] build upon the existing collective communication infrastructure, reuse these existing algorithms as submodules, and combine them to perform efficient and hierarchical collective operations. MagPIe [9] is another MPI system designed to construct collective operation trees in heterogeneous communication environments. MPICH2 [1] groups processes by nodes to limit the number of inter-node communication. Previous work [12] use trees that stratify the network deeper than two layers for further optimization. In work [8], it propose a topology-aware collective algorithms for large scale supercomputing systems that are tightly coupled through inter-connects such as InfiniBand. Work [2] propose a scalable network discovery service for InfiniBand available at the user level. With this service, it design a network-topology-aware MPI library to provide a topology aware mapping of the MPI processes. [5] address the application of pipelining to optimize a blocking All-reduce, for very large messages, of up to 1 GB, in the context of distributed DNN training on clusters of computer nodes. However, we utilize locality in Dragonfly interconnect topology to propose a topology-aware algorithm to minimize the communication latency in computing All-reduce.

6 Conclusion

The scale of deployment of supercomputers has increased year by year. These clusters provide a lot of computing resources for application developers. The demand for topology-aware collective operations is also increasing. This paper thus presented g-PAARD for use in Dragonfly systems. Essentially, g-PAARD designs each step of peer-to-peer communication to create a six-step process, regardless of node allocation. This ensures that at each step, each node communicates with its partner node through at most one hop. Therefore, gPAARD achieves the goal of minimum hops and short dependency chain more effectively than the previous All-reduce operation scheme. According to the simulation results, compared with the halving and doubling algorithm and the ring algorithm, the speed of g-PAARD is increased up to 4.33× and 287.34×. While, With today's supercomputer sizes and people relying more on group collectives, optimization of communication in group can be further researched. And, straggler processors is a potential performance bottleneck to collective operation, it is interesting to explore the sensitivity of straggler processors.

Acknowledgment. We thank the anonymous reviewers for their insightful comments. We gratefully acknowledge members of Tianhe interconnect group at NUDT for many inspiring conversations. The work was supported by the National Key R&D Program of China under Grant No. 2018YFB0204300, the Excellent Youth Foundation of Hunan Province (Dezun Dong), and the National Postdoctoeral Program for Innovative Talents under Grant No. BX20190091.

References

1. Hierarchical collectives in mpich2. Springer-Verlag (2009). https://doi.org/10. 1007/978-3-642-03770-2_41
2. Design of a scalable infiniband topology service to enable network-topology-aware placement of processes. In: SC 2012: Proceedings of the International Conference on High Performance Computing, Networking, Storage and Analysis (2013)
3. Archer, B.J., Vigil, B.M.: The trinity system. Technical report, Los Alamos National Lab. (LANL), Los Alamos, NM (United States) (2015)
4. Camarero, C., Vallejo, E., Beivide, R.: Topological characterization of hamming and dragonfly networks and its implications on routing. ACM Trans. Archit. Code Optim. (TACO) **11**(4), 1–25 (2014)
5. Castelló, A., Quintana-Ortí, E.S., Duato, J.: Accelerating distributed deep neural network training with pipelined MPI allreduce. Cluster Comput. **24**(4), 3797–3813 (2021). https://doi.org/10.1007/s10586-021-03370-9
6. De, K., et al.: Integration of panda workload management system with titan super-computer at OLCF. J. Phys. Conf. Ser. **664**, 092020. IOP Publishing (2015)
7. Faanes, G., et al.: Cray cascade: a scalable hpc system based on a dragonfly net-work. In: SC'12: Proceedings of the International Conference on High Performance Computing, Networking, Storage and Analysis. pp. 1–9. IEEE (2012)
8. Kandalla, K.C., Subramoni, H., Vishnu, A., Panda, D.K.: Designing topology-aware collective communication algorithms for large scale infiniband clusters: case studies with scatter and gather. In: IEEE International Symposium on Parallel & Distributed Processing (2010)
9. Kielmann, T., Hofman, R., Bal, H.E., Plaat, A., Bhoedjang, R.: MagPIe: MPI's collective communication operations for clustered wide area systems. In: ACM SIGPLAN Notices (1999)
10. Kim, J., Dally, W.J., Scott, S., Abts, D.: Technology-driven, highly-scalable drag-onfly topology. In: 2008 International Symposium on Computer Architecture, pp. 77–88. IEEE (2008)
11. Luo, X., et al.: HAN: a hierarchical autotuned collective communication framework. In: 2020 IEEE International Conference on Cluster Computing (CLUSTER), pp. 23–34 (2020). https://doi.org/10.1109/CLUSTER49012.2020.00013
12. Lusk, E., de Supinski, B.R., Gropp, W., Karonis, N.T., Bresnahan, J., Foster, I.: Exploiting hierarchy in parallel computer networks to optimize collective operation performance. In: Parallel and Distributed Processing Symposium, International, p. 377. IEEE Computer Society, Los Alamitos, CA, USA, May 2000. https://doi.org/ 10.1109/IPDPS.2000.846009
13. Rabenseifner, R.: Optimization of collective reduction operations. In: Bubak, M., van Albada, G.D., Sloot, P.M.A., Dongarra, J. (eds.) ICCS 2004. LNCS, vol. 3036, pp. 1–9. Springer, Heidelberg (2004). https://doi.org/10.1007/978-3-540-24685-5_1
14. Sensi, D., Girolamo, S., McMahon, K., Roweth, D., Hoefler, T.: An in-depth analy-sis of the slingshot interconnect. In: 2020 SC20: International Conference for High Performance Computing, Networking, Storage and Analysis (SC), pp. 481–494. IEEE Computer Society (2020)

15. Thakur, R., Gropp, W.D.: Improving the performance of collective operations in MPICH. In: Dongarra, J., Laforenza, D., Orlando, S. (eds.) EuroPVM/MPI 2003. LNCS, vol. 2840, pp. 257–267. Springer, Heidelberg (2003). https://doi.org/10.1007/978-3-540-39924-7_38
16. Thakur, R., Rabenseifner, R., Gropp, W.: Optimization of collective communication operations in MPICH. Int. J. High Perform. Comput. Appl. **19**(1), 49–66 (2005)

MPICC: Multi-Path INT-Based Congestion Control in Datacenter Networks

Guoyuan Yuan, Dezun Dong[(⊠)], Xingyun Qi, and Baokang Zhao

National University of Defense Technology, Changsha, China
{guoyuan,dong,qi_xingyun,bkzhao}@nudt.edu.cn

Abstract. Network congestion causes severe transmission performance loss. Congestion control is the key to achieve low latency, high bandwidth and high stability. Due to the high regularity of topologies, datacenter networks contain abundant resources of equal-cost paths inherently and have huge potential for high-bandwidth transmission. Taking in-network telemetry (INT) information as the congestion signal, congestion control algorithms can monitor the network based on precise link load data. Hence the INT-based algorithm is the focus of recent high-precision congestion control research. However, existing INT-based congestion control algorithms handle the single-path transmission between the sender and the receiver, which cannot make full use of the abundant equal-cost paths resources in the network. In this paper, we make the first attempt to propose a multi-path INT-based congestion control algorithm, called MPICC. Compared with the most advanced INT-based congestion control algorithm HPCC, our method realizes multi-path coordinated transmission of data packets in datacenter networks. We conduct extensive experiments under various workloads to evaluate the performance of MPICC. The experimental results show that under the condition of ensuring high throughput of end hosts, MPICC is able to use multi-path to transmit data packets, which greatly reduces the flow completion time (FCT) of the flow in the network. Specifically, under three common datacenter workloads, Cache Follower, Web Search and Web Server, MPICC reduces the average flow completion time of the network by 20.1%, 14.4%, and 39.5%, respectively, and shortens the 97th-percentile flow completion time by 39.9%, 18.9% and 57.9%, respectively.

Keywords: Congestion control · In-network telemetry (INT) · Multi-path · Datacenter networks · Programmable switch

1 Introduction

In order to meet diverse demands of customers, datacenter networks need to provide transmission services with high bandwidth, low latency and good stability. Congestion control is the key to achieve the above three characteristics.

© IFIP International Federation for Information Processing 2022
Published by Springer Nature Switzerland AG 2022
C. Cérin et al. (Eds.): NPC 2021, LNCS 13152, pp. 256–268, 2022.
https://doi.org/10.1007/978-3-030-93571-9_20

During recent years, with the in-network telemetry (INT) [3,4] information as the congestion signal, researchers design high-precision congestion control algorithms, which have greatly improved the datacenter's sensitivity to congestion and network transmission performance.

However, the prevailing INT-based congestion control algorithm solves the congestion of a single transmission path [1]. Actually, the datacenter network has abundant multi-path resources. Taking the classic Clos topology as an example, if two hosts from different racks need to exchange data packets, there are multiple paths through different core switches. Up to now existing INT-based congestion control algorithms do not carry out detailed research on multi-path transmission, which is a waste of rich path resources in the network.

Extending INT-based precise congestion control from single-path to multi-path enhances the utilization of network resources, which improves the transmission performance of the network and increases the stability of the network significantly [5,9].

To achieve multi-path INT-based congestion control, there are two important challenges.

First, the congestion control algorithm needs to handle multiple paths. On the one hand, the congestion control algorithm must manage each independent path. When a single path encounters congestion, the congestion control algorithm should be able to detect and alleviate it. On the other hand, the congestion control algorithm must coordinate data packets among multiple paths. For a flow, multiple transmission paths are existing concurrently, which increases the complexity of the congestion control algorithm. The sender needs to maintain the global sending behavior of each path, and select an appropriate path for the next data packet to be sent.

Second, the congestion control algorithm is supposed to cope with the situation where data packets arrive at the receiver out of order [5]. If the receiver does not receive the expected data packet, it sends a negative acknowledgement (NACK) packet to the sender, thereby triggering the sender's retransmission. A large number of data packets arriving out of order frequently trigger the retransmission, which greatly impairs network transmission performance.

The main contributions of this paper are as follows:

- We come up with a multi-path INT-based congestion control algorithm, called MPICC. MPICC utilizes INT information as the congestion signal and realizes multi-path transmission of flows in the network.
- We set temporary buffer for each flow at the receiver to store a certain number of out-of-order data packets. In this way, our method reduces the frequency of the sender's retransmission mechanism.
- We conduct extensive experiments to evaluate the performance of MPICC. Comparing with HPCC, experiments show the feasibility of MPICC.

2   Related Work and Motivation

2.1   The Need to Achieve Multi-path Congestion Control

Congestion is usually caused by a large number of data packets passing through a certain node concurrently during a short period of time. Congestion reduces network throughput, increases delay, and even triggers PFC pause frames, which seriously affects transmission performance. Therefore, congestion control is the key to adjust network transmission status and ensure the transmission performance [2]. The datacenter network has good multi-path characteristic, so designing congestion control algorithms that support multi-path management is of great significance to improve the transmission performance of the datacenter network [5,6].

Figure 1 shows a Clos topology which is commonly used in datacenter networks. It includes 4 core switches and 12 ToR switches, and each ToR switch is connected to 20 hosts. If $host_x$ and $host_y$ in Fig. 1 need to send data packets, there are four paths to choose in the network. In this paper, we take Clos as an example to discuss multi-path congestion control [13,14,17].

2.2   Challenges of Implementing Multi-path INT-Based Congestion Control

We find there are two main challenges in designing the multi-path INT-based congestion control algorithm [5,6]: one lies in the coordination of multiple paths when a flow is sending. The end host needs to have unified management of multiple paths and is supposed to select the optimal path for the subsequent data packets to be sent. The other is that the algorithm must be able to deal with the problem of out-of-order data packets at the receiver.

3   Design

3.1   MPICC Overview

Figure 2 is an example topology with 4 switches and 8 hosts. We show the architecture of MPICC on this topology. 4 switches are divided into two core switches *Core1* and *Core2*, and two top-of-rack switches (ToR switches) *ToR1* and *ToR2*. Hosts connected to *ToR1* is the sender, denoted as *s1*, *s2*, *s3*, and *s4*, and hosts connected to *ToR2* is the receiver, denoted as *r1*, *r2*, *r3*, and *r4*, respectively.

In this example, *s1* sends data to *r1*. Via different core switches, there are two different transmission paths between *s1* and *r1*, namely *s1-ToR1-Core1-ToR2-r1* and *s1-ToR1-Core2-ToR2-r1*.

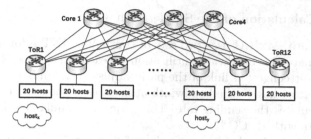

Fig. 1. A Clos topology with 4 core switches, 12 ToR switches and 240 hosts.

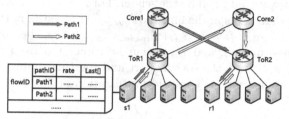

(a) An example topology with 4 switches and 8 hosts, where *s1* sends data to *r1*.

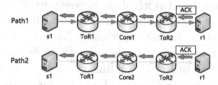

(b) The two paths between *s1* and *r1* in Fig. 2 (a).

Fig. 2. MPICC overview. In the topology shown in Fig. 2(a), *s1* sends data to *r1*. For each flow, under the management of MPICC, the sender records INT information of each path and maintains a multi-path list. The multi-path list records pathID, path sending rate and the last INT information of the path.

Taking the sending of a flow as an example, the network uses ECMP routing to send data packets through multiple paths. When the data packet passes through the switch on the path, it is marked with INT information, i.e., the path load data. When the data packet reaches the receiver, the receiver copies the INT information from the data packet into the ACK packet and sends it back to the sender. In this way, the sender is able to obtain the path information.

As shown in Fig. 2, for a flow, MPICC maintains a multi-path information list at the end host (sender). Each path has its own pathID. The pathID corresponds to the relevant information of the path, including the sending rate of the path, the normalized bandwidth utilization and the previous INT information. Whenever a new data packet is ready to be sent, the end host selects a relatively idle path (that is, the optimal path), and switches route the data packet according to the specified pathID.

3.2 Rate Calculation of the Single Path

For a single path, MPICC uses two sets of continuous INT information on the path to calculate the bandwidth utilization of each link. After that, MPICC selects the most congested link of the path, adjusts the sending window of the current path according to the bandwidth utilization of the most congested link, and then calculates the sending rate. This process is similar to the calculation of sending rate with HPCC [1].

To calculate the bandwidth utilization of link i, the key is to estimate the number of inflight bytes I_i of the link. Here inflight bytes are those that have been sent but not yet confirmed by the sender. For link i, I_i consists of two parts, including I_{link_i} that exists in the link and I_{queue_i} that exists in the queue of the current switch. There is $I_i = I_{link\_i} + I_{queue\_i}$. I_{queue_i} can be obtained directly from the INT information, and I_{link_i} can be calculated from the cumulative number of bytes sent from the specific port and the timestamp.

We use u_i to represent the bandwidth utilization of link i:

$$u_i = \frac{I_i}{bwth_i \times T} \tag{1}$$

Here $bwth_i$ represents the bandwidth of the egress port of the current switch, and T represents the basic RTT of the network. For the link bandwidth utilization, we set an expected value η. In the experiment, let η be 0.95, which means that we hope the bandwidth utilization of the network to be 95%. If $u_i > \eta$, it means that the actual bandwidth utilization of link i is higher than the expected value, there is a risk of congestion, and the sending rate needs to be slowed down. On the contrary, if $u_i < \eta$, it means that the link still has idle bandwidth and the sending rate should be appropriately accelerated.

For a single path, the sender selects the link with the highest bandwidth utilization and uses it to adjust the sending rate of the path. We let the bandwidth utilization of the most congested link be u_{max}.

In order to eliminate congestion quickly, when $u_{max} > \eta$, MPICC uses multiplicative transformation to adjust the sending window:

$$W_{(k_p)} = \frac{W_{(k_p)}}{u_{max(k_p)}/T} + W_{AI} \tag{2}$$

Here $W_{(k_p)}$ is the sending window of the current path p at sender k, $u_{max(k_p)}$ is the bandwidth utilization of the most congested link on the current path p, and W_{AI} is an additive parameter.

For better stability and convergence, when $u_{max} < \eta$, MPICC uses additive transformation to adjust the sending window:

$$W_{(k_p)} = W_{(k_p)} + W_{AI} \tag{3}$$

And multiple consecutive additive transformations trigger a multiplicative transformation shown in Eq. 2.

Since continuous INT information contains much repeated link status data, the reference window $W_{c(k_p)}$ is introduced in the calculation of the single-path sending window. After each change of $W_{c(k_p)}$, the sender records the sequence number seq of the next data packet sent on the path, and $W_{c(k_p)}$ does not change before the ACK packet corresponding to seq is returned to the sender. In this way, overreactions caused by repeated INT data are avoided. After introducing the reference window $W_{c(k_p)}$, the above Eq. 2 and Eq. 3 become:

$$W_{(k_p)} = \frac{W_{c(k_p)}}{u_{max(k_p)}/T} + W_{AI} \tag{4}$$

$$W_{(k_p)} = W_{c(k_p)} + W_{AI} \tag{5}$$

Algorithm 1 shows the process of calculating the sending rate of a single path.

3.3 Multi-path Coordination of MPICC

It is not natural for the sender to obtain the network multi-path information. It relies on the message in ACK/NACK packets to build the multi-path information list. Taking the sending of a flow as an example, the sender does not know any path status initially. The INT information piggybacked on returned ACK/NACK packets reflects the path data. By recording INT information from the coming ACK/NACK packets, the sender can obtain more and more path information, and further build a complete multi-path information list of the current flow.

In general, with MPICC, the transmission of data packets can be roughly divided into two stages. The first stage is the pathfinding stage under ECMP routing, and the second stage is the orderly transmission for data packets under the control of MPICC.

In the pathfinding stage, data packets passing through different paths transfer diverse path information to the sender by ACK/NACK packets. Each time an ACK/NACK packet containing new path information is received, the sender records and maintains the corresponding information. Since it takes at least one RTT from the sending of the first data packet to the receipt of the first ACK/NACK packet, data packets sent during the pathfinding stage is approximately one bandwidth-delay product (BDP) size.

After the pathfinding stage, the multi-path information list is established. Under the management of MPICC, the sender calculates the number of inflight bytes and the maximum allowable sending rate for each path. Whenever a new data packet needs to be sent, the sender adopts a path selection algorithm to specify a relatively idle path for the data packet. Then the data packet is routed in the network according to the specified path.

3.4   Out-of-Order Packets Control at the Receiver

Data packets are transmitted through multiple paths in the network concurrently. The congestion degree of each path is different, and the time for data packets of different path to reach the receiver is also different. This situation causes data packets arriving out of order at the receiver. If the expected data packet is not received for a long time, the receiver would send NACK packet, and then the sender's retransmission mechanism is triggered. However, retransmission greatly affects the performance of the network. Therefore, it is very important to control the out-of-order data packets at the receiver.

To increase the receiver's tolerance for out-of-order data packets, it is necessary to set a temporary buffer at the receiver. Compared with directly discarding unexpected packets, the temporary buffer can store several arriving packets. Whenever the receiver successfully receives the expected data packet pkt_i that arrives in order, where i represents the sequence number of the data packet. The receiver immediately goes to the buffer to check whether there is a data packet pkt_{i+1}. If such a data packet exists, the sender directly receives it from the temporary storage buffer. In this example, the method saves the overhead of retransmitting the data packet pkt_{i+1}. In this way, the receiver accepts a number

---

Algorithm 1. Sender algorithm. Compute the sending window and rate of a single path. *pkt* is the ACK/NACK packet; L represents the link feedback in the current ACK/NACK packet *pkt*; $L[i]$ is the INT of link i which has the following fields: *bwth* (the bandwidth of the egress port), TS (the timestamp), *txBytes* (the cumulative number of bytes sent from the current egress port) and *qLen* (the queue length of the egress port); *Last* represents the link feedback in the previous ACK/NACK packet of this flow; T represents the network base RTT; W represents the sending window; W_c represents the reference sending window.

---

procedure COMPUTEWINANDRATE(*pkt*)
    $u = 0$
    for each link i on the path **do**
        $txRate = \frac{pkt.L[i].txBytes - Last[i].txBytes}{pkt.L[i].TS - Last[i].TS}$
        $u' = \frac{min(pkt.L[i].qLen, Last[i].qLen)}{pkt.L[i].bwth \cdot T} + \frac{txRate}{pkt.L[i].bwth}$
        if $u' > u$ **then**
            $u = u'$
            $\tau = pkt.L[i].TS - Last[i].TS$
        end if
    end for
    $\tau = min(\tau, T)$
    $U = (1 - \frac{\tau}{T}) \cdot U + \frac{\tau}{T} \cdot U$
    if $U \geq \eta$ or $incStage \geq maxStage$ **then**
        $W = \frac{W_c}{U/\eta} + W_{ai}$
        if $pkt.seq > lastUpdateSeq$ **then**
            $incStage = 0$
            $W_c = W$
        end if
    else
        $W = W_c + View.W_{ai}$
        if $pkt.seq > lastUpdateSeq$ **then**
            $incStage + +$
            $W_c = W$
        end if
    end if
    $rate = W/T$
    $Last = pkt.L$
end procedure

Algorithm 2. Receiver algorithm. pkt_i represents the data packet with sequence number i. The receiver maintains a multi-path information list $pktList$ for each flow, which temporarily stores several arrived out-of-order data packets of the current flow. Each time the receiver successfully receives a data packet, it checks $pktList$ to find if there are subsequent data packets. If so, the receiver receives them directly from $pktList$. If $pktList$ is full, the receiver sends a NACK packet to the sender, triggering the sender's retransmission mechanism.

> **procedure** RECEIVEDATA(pkt_i)
>   **if** pkt_i is ordered **then**
>   send up pkt_i
>   **for** check $pktList$ **do**
>   **if** pkt_{i+1} is in $pktList$ **then**
>   Send up pkt_{i+1}
>   Remove pkt_{i+1} from $pktList$
>   check pkt_{i+2}, pkt_{i+3} ... until there is no ordered packet in $pktList$
>   **end if**
>   **end for**
>   **else**
>   **if** There is still room in $pktList$ **then**
>   store pkt_i in $pktList$
>   **else**
>   send $NACK$ to the sender
>   discard pkt_i
>   **end if**
>   **end if**
> **end procedure**

---

of out-of-order packets at the beginning, and only when the temporary buffer is full, the receiver sends the NACK packet to the sender, triggering the sender's retransmission mechanism.

The existence of the temporary buffer greatly improves the confirmation rate of data packets under the condition of multi-path transmission, and the number of NACK packets in the network is significantly reduced. Algorithm 2 shows the process of receiving data packets, checking the temporary buffer, and sending NACK packets at the receiver.

4 Implementation

4.1 The Format of INT Information

In order to use INT information as the congestion signal, the network needs switches that support INT functions. Each time the data packet passes through a switch, the switch marks the link load information of the current egress port at a specific field of the data packet. In this way, INT information arrives at the receiver with the data packet. After receiving the data packet, the receiver obtains the INT information of the complete path and adds it into a specific field of the ACK packet. Then the receiver transmits the ACK packet back to the sender.

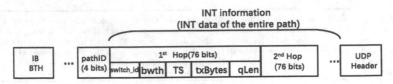

pathID (4 bits): It marks the specific path that the packet has passed through.
switch_id (12 bits): The source switch of the current link information.
bwth (4 bits): The bandwidth of the egress port.
TS (24bits): The timestamp when the packet leave the egress port.
txBytes (20 bits): The cumulative number of bytes sent from the egress port.
qLen (16bits): The queue length when the packet leaves the egress port.

Fig. 3. The format of INT information in the data packet and the ACK packet.

The format of the INT information in the data packet and the ACK packet is shown in Fig. 3. INT information is stored in a specific field which located between the IB BTH (Base Transport Header) and the UDP Header. The INT information required by MPICC mainly includes pathID and link load information of each hop. The latter includes the current egress port bandwidth $bwth$, the timestamp TS, the cumulative number of bytes sent by the current egress port $txBytes$, and the queue length $qLen$ of the switch when the data packet leaves the current port.

4.2   Implementation on RoCEv2 NIC (Network Interface Card)

On the OMNeT++ platform [11], we simulate the RoCEv2 NIC which supports MPICC. Figure 4 shows its protocol stack. Data packets from the upper application module first arrive at the OFA stack. After that, the MPICC module performs congestion control on the transmission of data packets. Then data packets reach the network protocol stack.

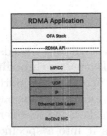

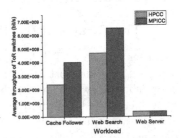

Fig. 4. Protocol stack of the RoCEv2 NIC.

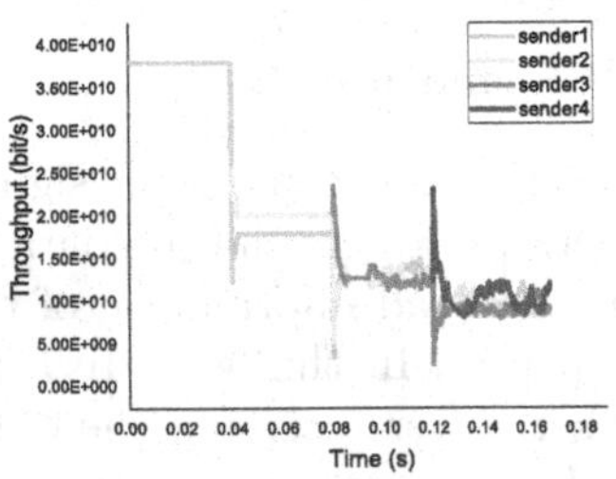

Fig. 5. The fairness of MPICC.

Fig. 6. The ToR switch average throughput of MPICC and HPCC under different workloads in the Clos topology.

5   Evaluation

5.1   Evaluation Setup

Experiment Platform. This work is carried out on the OMNeT++ 5.4.1 simulation platform with INET 4.0 framework [11].

Topology. We use the Clos topology shown in Fig. 1. It contains 4 core switches, 12 ToR switches and 240 hosts. Each core switch is connected to 12 ToR switches, and each ToR switch is connected to 20 hosts. In this experiment, we set the link bandwidth between the core switch and the ToR switch to 100 Gbps, and the link bandwidth between the ToR switch and the host to 10 Gbps. The delay between any two nodes in the network is 1 μs.

Workloads. Three common datacenter workloads are used: Cache Follower [8], Web Search [18] and Web Server [18].

Parameter Setting. When calculating the sending window with inflight bytes, parameters are set based on the paper [1]: $\eta = 0.95$, $maxStage = 5$, $W_{AI} = linkspeed * RTT * (1 - \eta)/n$. n represents the number of senders in the topology.

Performance Metrics. MPICC supports multi-path for data packets, which effectively improves the transmission efficiency of flows in the network. The main performance metric we focus on is the flow completion time (FCT).

5.2   The Fairness of MPICC

Through multiple senders sharing a path to send data to the same receiver, we test the fairness of MPICC. Specifically, we select 4 senders, *sender1*, *sender2*, *sender3* and *sender4*, which in turn send data at $t = 0$ s, $t = 0.04$ s, $t = 0.08$ s, and $t = 0.12$ s. The throughput changes of the senders are shown in Fig. 5. It can be seen that in addition to slight throughput jitter when a new flow join the network, MPICC exhibits good fairness. Here, the short-term jitter is due to the initial sending rate of a new flow.

5.3   MPICC Performance on Common Workload

To evaluate the performance of MPICC, we simulate a situation where two groups of hosts communicating with each other.

We first divide 240 hosts shown in Fig. 1 into two groups: 120 hosts connecting ToR switches *ToR1* to *ToR6* are regarded as group A, and the other 120 hosts connecting *ToR7* to *ToR12* as group B.

In our experiments, the two groups of hosts send data to each other randomly. We test the transmission performance of MPICC under the above three classic workloads, and the algorithm for comparison is HPCC.

Throughput in the Clos Topology. We record and calculate the ToR switch average throughput of HPCC and MPICC under three workloads on the Clos topology shown in Fig. 1. Figure 6 exhibits the experimental results. We can see that compared with HPCC, MPICC adopts multiple paths to transmit data packets, which significantly improves the throughput.

Flow Completion Time in the Clos Topology. From the perspective of the flow in the network, under the control of MPICC, multiple paths simultaneously transmit data packets, which improves the transmission efficiency of data packets and accordingly shortens the flow completion time in the network.

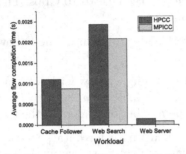

(a) The average flow completion time.

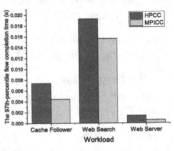

(b) The 97th-percentile flow completion time.

Fig. 7. The average and 97th-percentil flow completion time of MPICC and HPCC under different workloads in the Clos topology.

Figure 7 shows the average and 97th-percentile FCT of MPICC and HPCC under common datacenter workloads. We find that MPICC significantly reduce the flow completion time. Specifically, under Cache Follower [8], Web Search [18] and Web Server [18], MPICC reduces the average flow completion time of the network by 20.1%, 14.4%, and 39.5%, and shortens the 97th-percentile flow completion time by 39.9%, 18.9% and 57.9%, respectively.

It can be seen that through multi-path transmission of data packets, MPICC reduces the average flow completion time of the network, and the 97th-percentil flow completion time is also significantly reduced, indicating that the overall transmission efficiency of the flow in the network is significantly improved.

Figure 8 shows the FCT CDF diagrams of MPICC and HPCC under the three workloads. Due to the use of multi-path transmission, MPICC is able to shorten the overall flow completion time.

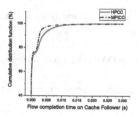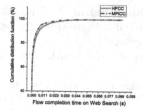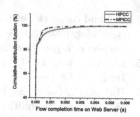

(a) CDF diagram of FCT under Cache Follower workload.

(b) CDF diagram of FCT under Web Search workload.

(c) CDF diagram of FCT under Web Server workload.

Fig. 8. The flow completion time CDF diagrams of MPICC and HPCC under different workloads in the Clos topology.

6 Discussion

The arrival of out-of-order data packets at the receiver is an inherent condition of multi-path, and the performance of MPICC is closely related to its degree. In future work, we would continue to study how to more efficiently alleviate the disorder problem and reduce the generation of NACK packets in the network.

7 Conclusion

In this work, we propose MPICC, which realizes multi-path INT-based congestion control of datacenter networks. On the one hand, by maintaining the multi-path information list at the sender, MPICC realizes single-path precise congestion control while taking into account the coordination of multiple paths, making full use of the rich equal-cost-path resources in datacenter networks. On the other hand, by setting the temporary buffer at the receiver, MPICC alleviates the impact of out-of-order data packets on network transmission efficiency. Simulation experiments under real workloads strongly confirm the effectiveness of MPICC in reducing network flow completion time.

Acknowledgment. We gratefully acknowledge the anonymous reviewers for their insightful comments and members of HiNA group at NUDT for their inspiring opinions. This work is supported by the National Key R&D Program of China under Grant No. 2018YFB0204300, Excellent Youth Foundation of Hunan Province (Dezun Dong) and National Postdoctoral Program for Innovative Talents under Grant No. BX20190091.

References

1. Li, Y., et al.: HPCC: high precision congestion control. In: SIGCOMM (2019)
2. Zhu, Y., et al.: Congestion control for large-scale RDMA deployments. In: SIGCOMM (2015)
3. In-band Network Telemetry in Broadcom Tomahawk 3 (2019). https://www.broadcom.com/company/news/product-releases/2372840

4. In-band Network Telemetry in Broadcom Trident3 (2019). https://www.broadcom. com/blog/new-trident-3-switch-delivers-smarterprogrammability-for-enterprise- and-service-provider-datacenters
5. Lu, Y., Chen, G., Li, B., et al.: Multi-path transport for RDMA in datacenters. In: NDSI (2018)
6. Raiciu, C., Barre, S., Pluntke, C., et al.: Improving datacenter performance and robustness with multipath TCP. In: SIGCOMM (2011)
7. Guo, C., Wu, H., Deng, Z., Soni, G., Ye, J., Padhye, J., Lipshteyn, M.: RDMA over commodity ethernet at scale. In: SIGCOMM (2016)
8. Roy, A., Zeng, H., Bagga, J., Porter, G., Snoeren, A.C.: Inside the social network's (datacenter) network. In: SIGCOMM (2015)
9. Huang, S., Dong, D., Bai, W.: Congestion control in high-speed lossless data center networks: a survey. Future Gener. Comput. Syst. **89**, 360–374 (2018)
10. Kandula, S., Sengupta, S., Greenberg, A., Patel, P., Chaiken, R.: The nature of data center traffic: measurements & analysis. In: SIGCOMM (2009)
11. OMNeT++ (2020). https://omnetpp.org/
12. Bai, W., Chen, K., Hu, S., Tan, K., Xiong, Y.: Congestion control for high-speed extremely shallow-buffered datacenter networks. In: APNet (2009)
13. Alizadeh, M., Kabbani, A., Edsall, T., Prabhakar, B., Vahdat, A., Yasuda, M.: Less is more: trading a little bandwidth for ultra-low latency in the data center. In: NSDI (2012)
14. Handley, M., Raiciu, C., Agache, A., Voinescu, A., Moore, A.W., Antichi, G., Wójcik, M.: Re-architecting datacenter networks and stacks for low latency and high performance. In: SIGCOMM (2017)
15. Hong, C.Y., Caesar, M., Godfrey, P.B.: Finishing flows quickly with preemptive scheduling. In: SIGCOMM (2012)
16. Alizadeh, M., Yang, S., Sharif, M., Katti, S., McKeown, N., Prabhakar, B., Shenker, S.: pFabric: minimal near-optimal datacenter transport. In: SIGCOMM (2013)
17. Greenberg, A., et al.: VL2: a scalable and flexible data center network. In: SIG- COMM (2009)
18. Alizadeh, M., et al.: Data Center TCP (DCTCP). In: SIGCOMM (2010)

Author Index

Printed in the United States
by Baker & Taylor Publisher Services